ARCHITECTURE FOR HEALTH AND WELL-BEING

A Sustainable Approach

ARCHITECTURE FOR HEALTH AND WELL-BEING

A Sustainable Approach

Edited by

María Eugenia Molar Orozco, PhD

AAP | APPLE ACADEMIC PRESS

First edition published 2023

Apple Academic Press Inc.
1265 Goldenrod Circle, NE,
Palm Bay, FL 32905 USA

760 Laurentian Drive, Unit 19,
Burlington, ON L7N 0A4, CANADA

CRC Press
6000 Broken Sound Parkway NW,
Suite 300, Boca Raton, FL 33487-2742 USA

4 Park Square, Milton Park,
Abingdon, Oxon, OX14 4RN UK

Apple Academic Press exclusively co-publishes with CRC Press, an imprint of Taylor & Francis Group, LLC

Library and Archives Canada Cataloguing in Publication

Title: Architecture for health and well-being : a sustainable approach / edited by María Eugenia Molar Orozco, PhD.
Names: Molar Orozco, María Eugenia, editor.
Description: First edition. | Includes bibliographical references and index.
Identifiers: Canadiana (print) 20220244782 | Canadiana (ebook) 20220244847 | ISBN 9781774910122 (hardcover) | ISBN 9781774910139 (softcover) | ISBN 9781003282402 (ebook)
Subjects: LCSH: Sustainable architecture. | LCSH: Architecture—Health aspects. | LCSH: Architecture—Human factors.
Classification: LCC NA2542.36 .A73 2023 | DDC 720/.47—dc23

Library of Congress Cataloging-in-Publication Data

CIP data on file with US Library of Congress

ISBN: 978-1-77491-012-2 (hbk)
ISBN: 978-1-77491-013-9 (pbk)
ISBN: 978-1-00328-240-2 (ebk)

About the Editor

María Eugenia Molar Orozco, PhD
Full Professor, School of Architecture, Autonomous University of Coahuila, México

María Eugenia Molar Orozco, PhD, is a Full Professor at the School of Architecture at the Autonomous University of Coahuila, México. She is a member of the National System of Researchers (level 1). She has experience in research on thermal comfort and sustainability and the comfort of indoor and outdoor spaces. She is the author of three books with prestigious publishers, several journal articles, and several book chapters. She has been a thesis director of four MS theses and 31 BSc theses and has developed more than four research projects. In 1990, Dr. Molar received a degree in Architecture, and in 2002, she received a master's degree in Architecture, and in 2009, she received a PhD in Architecture.

About the Chapter Authors

Gonzalo Bojórquez Morales, PhD, is a Research Professor in the Faculty of Architecture and Design at the Universidad Autónoma de Baja California, Mexico, where he is responsible for the Environmental Design Laboratory. He has research experience in bioclimatic architecture and thermal comfort and habitability and energy in architecture. He is the author of book chapters and several articles in indexed journals. He is a member of the National System of Researchers (level 1). He received his master's degree and PhD in Architecture.

Rubén Salvador Roux Gutierrez, PhD, is a Research Professor at the Institute of Higher Studies of Tamaulipas, Mexico. His research focuses on earth construction and habitability. He is the author of several books from prestigious publishers as well as book chapters and articles in indexed journals. He is a member of the National System of Researchers (level 1). He received a master's degree in Engineering and his PhD in Architecture.

Rolando-Arturo Cubillos-González, PhD, is Research Professor in the Faculty of Architecture at the Universidad Católica de Colombia, Mexico. His research is focused on resilience design. He is the author of one book by a prestigious publisher as well as book chapters and articles in indexed journals. For several years, he was director of the research group Sustainability, Environment and Technology. He received a master's degree in science, focusing on habitat (the study of the structure, function and change of human habitat, and the ways in which humans react with this habitat; specialization in social housing complexes), and he is currently a PhD candidate in Technology and Innovation Management.

Juan Flavio Molar Orozco is a Professor at the Tecnológico Nacional de México/Campus Tecnológico de Ciudad Madero, Mexico, where he is Head of the Industrial Engineering Department. He is the author of three book chapters. He is a logistics specialist. He holds a master's degree in industrial systems engineering.

Contents

Contributors

Gonzalo Bojórquez-Morales
Faculty of Architecture and Design, Autonomous University of Baja California, Blvd Benito, Juarez S/N, University Unit, C.P. 21280, Mexicali, Baja California, México,
Mobile: +(52)6865664250, Ext.117, E-mail: gonzalobojorquez@uabc.edu.mx

Rolando Arturo Cubillos González
Faculty of Desing, Catholic University of Colombia, Diagonal 46 A # 15 B – 10, Bogotá, Colombia,
Mobile: +(57)13277300, Ext. 3109, E-mail: racubillos@ucatolica.edu.co

Rubén Salvador Roux Gutiérrez
Faculty of Architecture, Institute of Higher Studies of Tamaulipas, Ave. Dr. Burton E. Grossman 501 Pte.Col Tampico-Altamira Sector 1 C.P. 89605, Tampico, Tamaulipas, México,
Mobile: +(52)8332302550, E-mail: ruben.roux@iest.edu.mx

Juan Flavio Molar Orozco
Industrial Engineering, National Technology of Mexico/Technological Campus of Ciudad Madero, Av. 1o. de Mayo s/n (Sor Juana Inés de la Cruz, Col. Los Mangos) 89440, Ciudad Madero, Tamaulipas, México, Mobile: +(52)8333574820, Ext. 3060, E-mail: juan.mo@cdmadero.tecnm.mx

María Eugenia Molar Orozco
Faculty of Architecture, Institute of Higher Studies of Tamaulipas, Ave. Dr. Burton E. Grossman 501 Pte.Col Tampico-Altamira Sector 1 C.P. 89605, Tampico, Tamaulipas, México,
Mobile: +(52)844 8691001, Ext. 8, E-mail: mariamolar@uadec.edu.mx

Abbreviations

CO_2	carbon dioxide
COVID-19	coronavirus disease 2019
DBT	dry bulb temperature
EPT	equivalent physiological temperature
GDP	gross domestic product
GEA	Geobiological Studies Association
IBT	internal body temperature
MST	mean skin temperature
MTSI	means by thermal sensation interval
NO2	nitrogen dioxide
PHEIC	Public Health Emergency of International Concern
PWD	people with disabilities
RUROS	rediscovering the urban realm and open spaces
SD	standard deviation
SO_2	sulfur dioxide
WC	water closet
WHO	World Health Organization

Foreword

Writing in 2020, one cannot but reflect on the implications for the architecture of the coronavirus pandemic. As we look back on a year during which we were shut-in and our homes became our offices and schools, we see how we have collectively examined these spaces and perhaps understood more clearly than ever the importance of the built environment on our health, both mental and physical. In architecture circles, this has been a year of debates on the city and its buildings in the light of the global pandemic. We have posed profound questions as to what we will learn and how our homes, offices, schools, and public spaces will be transformed. We have circulated videos showing examples of modern architecture of the 20s—hospitals, sanatoriums, and schools—that illustrated the changes that came about as a result of the flu epidemic of 1918 and the constant threat of tuberculosis. As these examples circulated, it became apparent that what had been learned had also been forgotten. The open classrooms that allowed for constant ventilation had succumbed over the years to more traditional closed solutions.

Today we, as architects, face two major challenges, one related to health, and specifically the pandemic, and the other related to climate change; both with major repercussions on the well-being of the inhabitants of the buildings we design. While wealthy countries devise expensive solutions, poorer nations must rely on creativity, ingenuity, and adaptability to compensate for the lack of financial resources. In the current panorama of great needs with few resources, applied research has become a priority in Mexico, research that focuses on the solution of specific problems in order to make real contributions to the welfare of the population. These solutions, in turn, have to rely on appropriate technology and be attentive to local conditions in order to be sustainable and successful in the long term.

Mexico is the host to diverse climates and ecosystems. In spite of stereotypes that associate the country with its Northern deserts, vast portions of the country host other ecosystems. The central highlands are characterized by pine forests, and many of the coastal areas and lowlands by dense tropical forests. This diversity has spawned the formation of research groups in Mexico's public universities. Research into thermal comfort has branched out into the design of passive strategies and technological innovation—often proposed for mass use in the large low-income housing developments that appear all over the country—that can mitigate the effects of intense dry heat in the north or the tropical conditions of the south. Not only is sustainability an issue in a moral sense, but it is also a necessity for low-income families who cannot afford the cost of artificial air conditioning. Another area of applied research in Mexico explores alternative materials, often related to local building traditions and natural materials associated with regional architectures. This has been particularly fruitful in the development of new formulas for earthen structures and for the use of palms.

All three of these areas, presented in this volume, provide the reader with recent research done in Mexican universities focused on the intersection between well-being, health, and architecture. In this volume, young as well as experienced researchers make integral contributions that reflect on building, spatial qualities, materials, and comfort in order to ensure a healthier environment in housing in their region.

—Catherine Ettinger
Morelia, Michoacán

Preface

The aim of the book is to analyze the importance of architecture, not only from the perspective of a beautifully realized construction, but to show some aspects that are contemplated to achieve the true ultimate goal, which is the well-being of the user, involving the outdoor and indoor spaces that together are related to obtaining a healthy environment, considering aspects of design, materials, environmental parameters, and their final process.

The book provides another perspective on architecture, which now more than ever due to the COVID-19 paradigm requires a return to the role of welfare promoter, which breaks the paradigm of current architecture, and returns to the main root of construction, which is to give security and welfare to those who inhabit it. The wording was developed during the pandemic, considering everything seen and known, without leaving aside other aspects of housing that affect health.

Translated with www.DeepL.com/Translator (free version), each chapter will explain from different angles a topic that allows direct actions for the benefit of the human being to achieve a better quality of life within constructions. The first part shows an overview of the situation and problem, both the context and inside of the construction. The second part sets out the role of the urban aspect in relation to health with alternatives to consider. The third chapter presents alternative materials that could minimize the existence of the virus and bacteria inside the construction. The next section highlights the role of resilience in a healthy architecture, and finally, it highlights the importance of adequate logistics in order to achieve the goal of a healthy architecture.

The contribution of this book is to provide basic information and generate a change in attitude in those who develop this profession but also to re-enter the role of the architect for the well-being of people who use the spaces, eradicating the negative idea that has been generated over time, about the contribution of an architect in designing spaces, because of bad practice of some. Certainly, this book will be an important asset in teaching activities about applications of diverse resources in architecture and for researchers in the aforementioned architecture domains as well as for professionals in the construction industry.

Finally, we thank all those who directly or indirectly inspired and supported the development of the book.

— ***María Eugenia Molar Orozco, PhD***

CHAPTER 1

The Role of Architecture to Achieve Well-Being

MARÍA EUGENIA MOLAR OROZCO

Faculty of Architecture, Autonomous University of Coahuila, Boulevard Fundadores Km 13, City University, Arteaga Coahuila CP – 25354, Arteaga, Coahuila, Mexico, E-mail: mariamolar@uadec.edu.mx

1.1 INTRODUCTION

Over time, architecture has been debunked. People do not consult an architect to design their homes. They turn to the builders, and some are in the hands of accountants or other professionals who don't have the criteria for decision-making in design. Even if they have architects at their service, they cannot express their opinion. What predominates is the gain even at the expense of the customers; only when they can pay more, they can have a home to their liking according to their particular needs, making some changes to a construction that is already on the market. A space is not only made up of measures or dimensions, it also considers quantitative and qualitative aspects like lighting, sound, temperature, humidity, wind, materials, pollution, natural phenomena, and even psychological aspects, and considering all these parameters in the design process is important for a healthy and dignified space.

The United Nations Committee on Economic, Social, and Cultural Rights in the Committee's General N°4 (1999) Comment on the Right to Adequate Housing says:

> *The concept of "adequate housing" should provide more than four walls and a ceiling. It means having an accessible place (for vulnerable people),*

Architecture for Health and Well-Being: A Sustainable Approach. Maria Eugenia Molar Orozco, PhD (Ed.)

protection from adequate health, moisture, space, safety, lighting, and ventilation risks, adequate location in relation to basic work and services, availability of basic infrastructure, and at a reasonable cost (UN HABITAT, 2010).

At the end of 2019, a coronavirus was identified in Wuhan, China, and until March 11th, 2020, was recognized as a global pandemic by the World Health Organization (WHO), creating a new paradigm in 2020 about architecture, which is not something new.

In every pandemic, like the 1918 influenza, existing problems become visible, as many experts point out, like Leilani Farha from UN-Habitat, and Emilia Saiz, Secretary-General of UCLG, and authorities from other countries, and researchers from different universities.

The architecture historian Paul Overy, highlighted that the primordial elements in architecture are light, ventilation, and open spaces, and in hospitals and houses, those elements should be based on three axes: good air circulation, natural light, and easy-to-clean surfaces (Ducan, 2020); something that has been lost over time.

This chapter will explain the reasons for taking these aspects into account in the design, explaining, and analyzing the morphological effects that affect the interior of the buildings.

1.2 HEALTHY ENVIRONMENT

In the second half of the 20th century, two aspects related to environmental problems became visible. The green one focused on the negative effects of human activity on the natural environment and sustainability; and the blue one, concerned about the effects of the environment on the health and well-being of humanity, known as environmental health. In 1993, the WHO defined as Environmental Health "*those aspects of human health, including quality of life, determined by physical, chemical, biological, social, and psychosocial environmental factors; both in theory and in practice*" (cited by Ordoñez, 2000).

Some authors such as Bueno (1998), organizations like GEA Association for Geobiological Studies and UN-Habitat, and some companies like Siber, express the concept of a healthy environment, and a healthy home.

It is important agreeing on the importance of health through of a construction carried out conscientiously, which involves not only the building but also its location regarding the environment, the materials and other elements that are combined to create a home, having a product that provides a healthy habitat for users?

It could be said that the house is a living being that if not done properly can cause diseases. Then, the following question arises from the above, what is a healthy environment?

A space that promotes health and well-being is a healthy environment, in which good circulation and renewal of the air, zero noise pollution, and avoiding conditions that allow macrobiotic agents to proliferate in the space must be considered. These pollutants not only affect the human being, but the building itself, generating bad odors, deteriorating construction materials, leading to a vicious circle. Although this may sound like a fallacy, because sometimes it is not possible to be 100% free of any pollutant or it may be beyond our control, the goal of a home will always be where you want to get to at the end of the day.

For this, it is necessary to avoid the use of synthetic materials, supplying sufficient fresh air, and solar radiation. With the goal of having a healthy or not so harmful environment, either for a new home or one that requires renovation, also considering other agents. From the geobiology point of view, the following are considered:

1. Natural Factors: Telluric and geophysical alterations, Hartmann lines, curry lines, vortices or cosmoteluric chimneys, radioactivity, and radon gas.
2. Artificial Factors: Electrical and electromagnetic pollution, high frequency, mobile telephony and wireless networks, air quality or habitat chemistry, lighting (GEA, 2020).

The construction must contemplate the criteria of health and harmony, not only aspects of design and construction, the goal is that the inhabitants feel comfortable and maintain their well-being. For this, the materials' selection used in constructions is important and the materials that are not polluting or affecting their health (Figures 1.1 and 1.2).

Most of the time, humans are in a closed space, sometimes hermetically, without the possibility of natural ventilation, only supported by a mechanical system to circulate the air, and in the majority of these cases, a good air renewal is not achieved, as the air itself is stale. According to Espinosa (2020), "biohabitability analyzes the quality of a space to be inhabited. It is a science that studies the influence of the indoor environment on people's health and well-being." Therefore, based on the set of parameters existing in the habitat and its construction, it generates a favorable environment for humans, their well-being and health.

FIGURE 1.1 Local materials made of earth in 2005.
Source: Photographs of the author.

FIGURE 1.2 Local materials made of wood and stone in 2006.
Source: Photographs of the author.

A healthy or bio-habitable habitat is characterized by:

- Natural lighting;
- Indoor temperature, humidity, and acoustic conditions must be within upper and lower comfort limits;
- Not emitting pollutants of physical, chemical, or biological origin;
- Not emitting electromagnetic radiation;
- And not receiving radiation from the outside (either natural radioactivity or electromagnetic fields of artificial origin).

An unhealthy habitat causes:

- Fatigue, tiredness, and insomnia;
- Headache or migraine;
- Frequent infections;
- Allergies and sensitivity.

These effects are part of the sick building syndrome, to identify it, a follow-up must be surveyed to see if it repeats constantly and in one or more specific spaces, but biohabitability encompasses more physical, chemical, and biological risk factors, depending on the context and geographic location. The best way to ensure a healthy environment is to carry out a biohabitability study according to the SBM-2015 Technical Standard and the Supplement "Framework conditions for technical measurements, clarifications, and complements."

The Spanish Association of Geobiological Studies has a decalogue with 10 aspects and considerations when designing, based on biohabitability criteria (own elaboration based on Bueno (1998)):

1. Housing and surroundings. Select a place that is not vulnerable, that is environmentally friendly, and that the construction is adapted to the environment, having green areas for a good microclimate that generate comfort conditions inside the spaces.
2. Global assessment of risk factors around the home: avoid noise sources and olfactory pollution that do not allow a healthy space.
3. A geobiological study of the site to locate intimate and social spaces, and even work spaces where they can be affected by terrestrial radiation, seismicity, energy line crossings and maintain a healthy space.
4. Avoid electromagnetic contamination both from the outside (proximity to power towers) and inside due to poorly executed or adequate installations.

5. Bioconstruction criteria based on a diagnosis, selecting the appropriate materials for the climate to achieve a comfortable, healthy, and ecological space with the support of clean and renewable energy.
6. Select healthy and ecological materials (not with a closed life cycle) with zero emissions of pollutants to ensure the health of the user who inhabits it.
7. Indoor air quality: ensure adequate renewal to avoid stagnation and maintain good air quality inside the space.
8. Natural lighting: analysis of the place on elements that may harm or obstruct the entry of the sunlight, place the openings in such a way that natural light can be used to demand less energy consumption.
9. Optimal management of natural resources: include alternatives to recycle, separate, and contain waste that can be processed as compost for gardens, if necessary, besides having water recycling facilities, which it is hard to achieve.
10. Responsibility of the home with life, health, and the environment: a construction that has low environmental impact from its construction to its demolition.

A house is a protective entity for the people who live in it. This construction interacts with the outside, being a connector with the inside, and transfers energy according to environmental conditions. For Bueno (1998), a healthy home must contemplate:

- Site and environmental assessment;
- Correct orientation and use of passive energies;
- Harmonious forms integrated into the landscape or local architecture;
- Construction material, healthy, non-toxic, or radioactive;
- Construction systems;
- Thermal comfort (heating, cooling, and insulation);
- Acoustic comfort;
- Harmony of colors and decoration (light and color);
- Electrical installation;
- Air quality.

These aspects are sometimes omitted due to cost issues. The phrase of Bueno (1998) is interesting, and now more than ever, "*tell me where you live, and I'll tell you what you suffer*." Eventually, extremely, or poorly understood technology can affect whoever lives in a home by making the mistake of constructing a hermetic structure with the justification of the use of mechanical systems, but when these fail, comfort conditions are lost to the not having alternatives to help defray the situation.

Building in vulnerable areas will always generate stress for the inhabitant, as Bueno (1998) points out, housing is part of the ecosystem, since it demands energy and generates waste. It is an entity that breathes and interacts through its envelope, everything in the environment directly affects or benefits the construction.

The first factor to consider, is the land should not be located in a vulnerable area. Areas with those characteristic are purchased by construction companies for be cheaper and that customers are not aware of or are built without taking these factors into account (Figures 1.3 and 1.4).

FIGURE 1.3 Houses near the train tracks.
Source: Photograph of the author (2020).

In Mexico, when a house is purchased, most owners do not considered check if buildings have any problems, in Spain, there is a basic quality requirement in a construction, and owners must have a certificate that validates this.

1.3 THE ROLE OF ARCHITECTURE IN THE FACE OF PANDEMICS

Through history, each pandemic has affected humanity by modifying what was already established in architecture and even gave ideas for new furniture proposals to improve the quality of life inside spaces.

FIGURE 1.4 Damage of a house from being near a train track.
Source: Photograph of the author.

For example, is the water closet (WC) or toilet, played an important role along with other measures for the physical and emotional health of the user.

The first documented toilet similar to the ones we use today, with a drainage channel, cistern and bowl, is from 4,000 years ago, in the royal palace of Knossos on the island of Crete.

Rome, around its first century, had facilities to unload the bladder; these were public urinals called urinal columns.

Roman ordinances of that time prohibited "*dirtying on stairs, hallways, or closets; on the walls of palaces, churches or houses for public use,*" as recounted by the 16th century Toledo historian, Father Juan de Mariana (CurioSfera Historia, 2019).

In Mexico can find vestiges of Muslim-style toilets from the Mayan culture in Chiapas (Figures 1.5 and 1.6).

FIGURE 1.5 A space for a toilet, 2019.
Source: Photographs of the author.

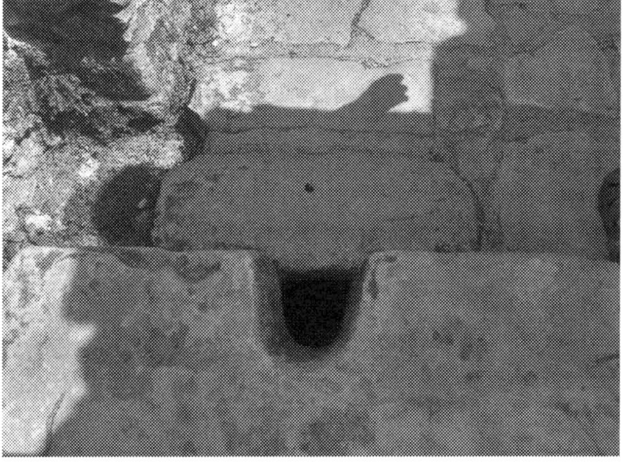

FIGURE 1.6 A close-up view of the toilet, 2019.
Source: Photographs of the author.

In 1596, Sir John Harrington, godson of Queen Elizabeth I, conceived a toilet connected to a water tank that carried waste when flushed. The design was thought for the queen who she did not like going to the bathroom because it caused her stress, but it was not well received, some point out that it was due to the absence of sewage networks (Sánchez, 2019) or the refusal to accept something new. The custom was to use the potty (high and cylindrical urinal), shouting "water goes!" emptying the potty into the streets, generating a focus of sanitary infection since the feces brought diseases such as typhus (CurioSfera Historia, 2019).

In 1775, Alexander Cummings took up the idea and invented the first modern toilet. He patented a toilet with the same principle as Harrington, the great innovation was that the drain was done through a siphon, an "S" shaped pipe that allows the liquid level in the bowl to be maintained, creating a clean water barrier that prevents bad odors from returning to the toilet, with the possibility of installing the toilet in a home without problems. In 1819, Albert Giblin created a model very similar to the current ones, without a valve in the bowl. In 1849, Thomas Twyford made the first ceramic toilets, and in 1880, Thomas Crapper, acquired the Giblin patent and invented the float, which serves to automatically close the water flow in the cistern (Sánchez, 2019).

The significance of the WC was such that it was considered in the act of the British Parliament of 1848 that forced the installation of toilets in new homes, although decades passed before it reached all houses (Sánchez, 2019), even when the sewer system in London worked until 1860 (CurioSfera Historia, 2019). In Mexico there are still some houses with a bathroom outside (Figure 1.7), however, this invention was crucial to save lives through hygiene practices, making the WC an important part of a construction (Figure 1.8).

1.3.1 PLAGUE

During the 14th and 18th centuries, the Black Death was one of the most lethal plagues in the history of mankind (Figure 1.9), caused by the bacterium *Yersinia pestis*, it ravaged various areas of the planet. In history there are three types of pests registered:

- Plague of Justinian from the 6th century;
- Black plague in the middle of the 14th century; and
- The third major plague pandemic originated in the Chinese province of Yunnan in 1855 (Huguet, 2020).

FIGURE 1.7 A bathroom in a vulnerable home 2015.
Source: Photograph of the author.

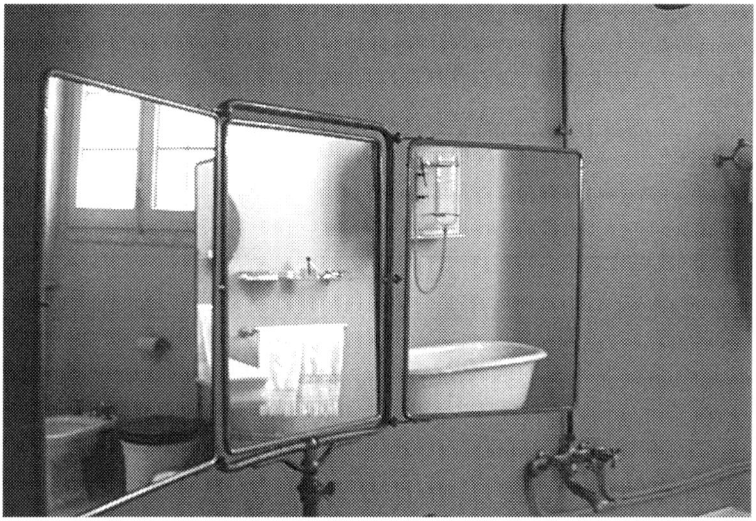

FIGURE 1.8 The interior of a bathroom in Gaudí's modernist era, 2003.
Source: Photograph of the author.

The recurring element in all three is hygiene and avoiding contact with the sick. Architects cannot be in the comfort zone, which has been allowed for a long time, in history there have been factors or variables that have demanded creativity and have drawn the shape of cities. Pandemics have always generated a change in all fields, in the social and cultural habits, and customs, and even in the spaces.

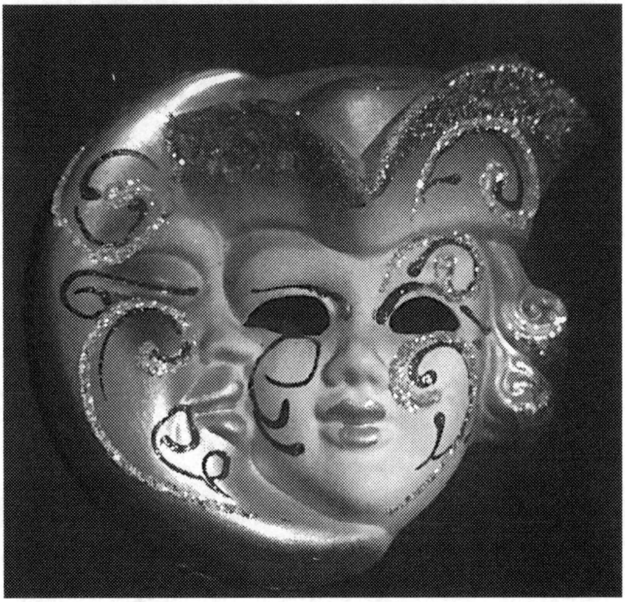

FIGURE 1.9 A mask based on those used by 17th-century doctors.
Source: Photograph of the author.

1.3.2 TYPHUS

It is a bacterial disease spread by lice or fleas (Medline plus, 2020). The first description of the disease is from 1489 in Spain during the Nasrid kingdom of Granada. Between 1577 and 1759 "*Gaol fever*" or "*Aryotitus fever*" (Aryotus fever) was common in English prisons, and it is believed that it was typhus. Typhus was caused by the overcrowding of prisoners in dark and poorly cleaned cells, facilitating the spread of lice, and outbreaks routinely appeared in Europe from the 16th to the 19th centuries (Sánchez, 2014; Trials of War Criminals, 1949).

The first strategy to avoid contracting typhus is hygiene, this was applied during the Second World War in a Warsaw ghetto, in Poland. In 1940, poor sanitation, famine, and a population density of 5 to 10 times larger than any city, it was the perfect breeding ground for the typhus epidemic to spread like wildfire. At the end of 1941 the contagion curve flattened, due to the strategies implemented by the Jewish doctors, followed among the inhabitants of the ghetto which were social distancing, general hygiene and clean spaces were promoted (Chaparro, 2020).

1.3.3 CHOLERA

In the 19th century, choleras' effect was devastating, shocking Asia and Europe. The world was beginning to globalize, and international business transactions moved huge amounts of money. The bacterium producing cholera is quite common in areas where there is no drinking water and poor hygienic conditions. Water is the main source of contagion and it is transmitted between people. The lack of hygiene and overcrowding among workers helped in its spread (Ríos, 2020). Unhealthy cities and the lack of sewerage were the main causes for this disease (France 24, 2020), spreading through an infected well, giving way to legislative reforms and sanitary infrastructure (Confidential, 2020), and changing the cities.

This generated a change at the urban level and cities planning, including drinking water treatment, connection to the public sewer and to a septic tank, siphon latrines, simple pit latrines and improved and ventilated latrines (González, Casanova, and Pérez, 2011).

Another change in the urban aspect is the intervention of Ildenfonso Cerdá in Barcelona, Spain, by introducing his draft for the *Ensanche*, he pointed out that, when cholera disappears, it leaves a mark in each house due to the poor hygienic conditions of the villages Barcelona and the overcrowding of inhabitants on a small surface. Cerdá's main work was based on detailed analyzes of the climate, such as temperature, wind, and air purity, up to the detailed quantification of the surfaces and volumes of the various types of buildings regarding the needs of the occupants (Laboratori D'urbanisme, 1992).

Cerdá suggested that in the simplest and most hygienic of all the combinations between houses, there should be a space equal to the area of a house (Figure 1.10) to receive light, air, and ventilation from all sides without requiring patios, which are perennial reservoirs of stale air (Laboratori D'urbanisme, 1992). The shape of an urban design should have a reticular expansion, the orientation of the vertices coinciding with the cardinal points, with the objective that most of the sides receive sunlight (Figure 1.11).

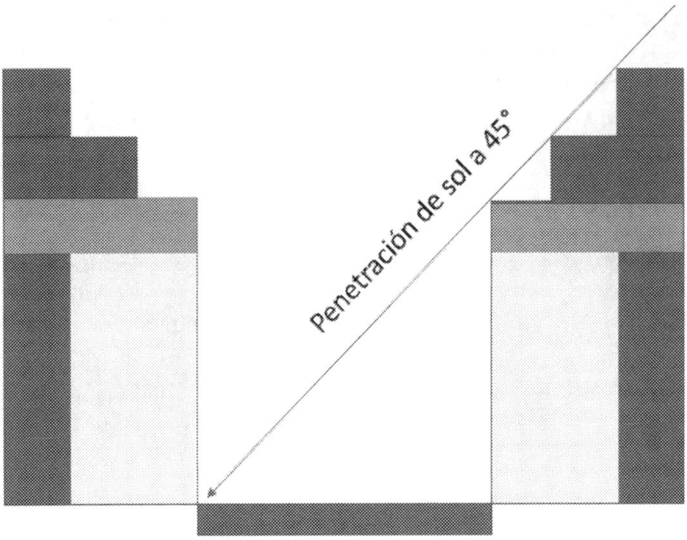

FIGURE 1.10 Height and space relationship for sun access.
Source: Own elaboration.

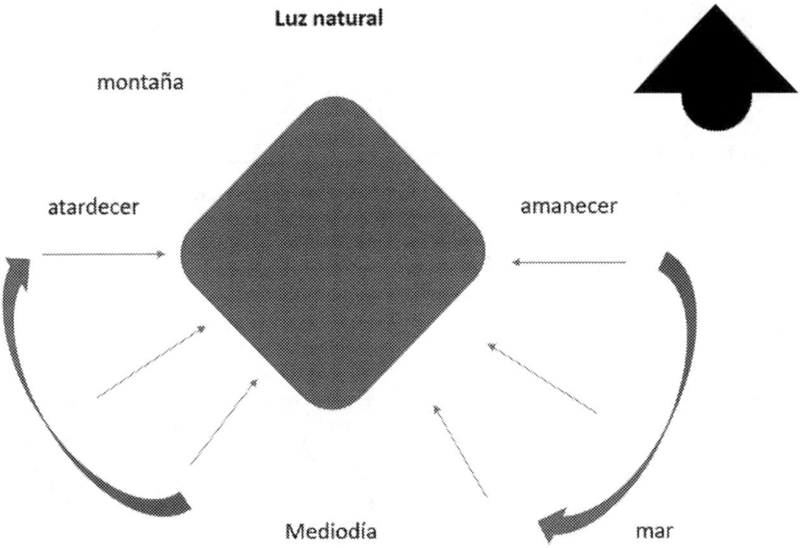

FIGURE 1.11 Orientation of the blocks in the Ensanche of Barcelona.
Source: Own elaboration.

1.3.4 MEASLES

Measles has been known for centuries, the first descriptions are attributed to the Hebrew physician Allyehudi in the 7th century and in the 10th century to the Persian physician, Rhazes, who named it "rash." In the 17th century, during a severe epidemic in London, measles, and smallpox were reported, and people thought were one disease (Fadic and Repetto, 2019).

In the 21st century, measles is no longer a well-known disease, but still requires isolation of the sick person, personal hygiene and keeping clean surfaces.

1.3.5 INFLUENZA

The influenza or the flu pandemic of 1918 during World War I was spread in confined and closed spaces, and the high density of troops helped to spread the disease (CDC, 2018). Based on this, there were architects who generated new proposals, such as Édouard Jeanneret known as Le Corbusier who revolutionized architecture thanks to the reflections that were born during his days of confinement in his Parisian department to survive the Spanish influenza epidemic.

During the following years, Le Corbusier became obsessed with the relationship between space and disease, between 1920 and 1921 he published articles in the magazine L'Esprit Nouveau, where he proposed new ways that architecture should respond, based on a fresh vision on hygiene in the cities. He and other architects were concerned about the impact of architecture on public health. The architecture historian Paul Overy highlighted light, ventilation, and open spaces as essential elements in modern architecture, mainly in hospitals and homes, based on three conceptual axes: good air circulation, natural light, and surfaces easy to clean (Ducan, 2020); something that has been lost in the 20th and 21st centuries in some constructions.

The proposals can still be seen in hospitals, because the bacteria survived in dark and dusty spaces, the sun and air were the weapons to help reduce this, the first architectural aspect was the use of wide windows or walls that could be moved in order to have good ventilation (Figures 1.12–1.15). Hospitals not only had large windows but also terraces where patients could take the sun and fresh air to recover from the disease (Figures 1.16 and 1.17), in the houses, architects used columns so the rooms were far from the surface, the modernist architecture and minimalism were born with the sole objective of having surfaces easy to clean, that did not keep dust.

FIGURE 1.12 Windows in the Civil Hospital of Tampico, Tamaulipas, 2012.
Source: Photographs of the author.

FIGURE 1.13 High ceilings in the Civil Hospital of Tampico, Tamaulipas, 2012.
Source: Photographs of the author.

FIGURE 1.14 The Civil Hospital of Tampico, Tamaulipas, 2012.
Source: Photographs of the author.

FIGURE 1.15 Interior doors and patios of the Civil Hospital of Tampico, Tamaulipas, 2012.
Source: Photographs of the author.

FIGURE 1.16 Terrace for patients of the Civil Hospital of Tampico, Tamaulipas, 2012.
Source: Photographs of the author.

FIGURE 1.17 The view of the terrace for patients of the Civil Hospital of Tampico, Tamaulipas, 2012.
Source: Photographs of the author.

This strategy was even applied in schools called open air schools or anti-tuberculosis schools, an example that still exists is the Open-Air School (*Openluchtschool*) which was constructed between 1927 and 1930, located in Cliostraat 36–40, Amsterdam, The Netherlands, by the architects Johannes Duiker and Bernard Bijvoet (Figure 1.18). The fundamental condition of the design is that all the classrooms were arranged, so they received the maximum amount of light and sun, and that the terraces worked for two classrooms, even in bad weather, since they were protected by covers and sheltered from the wind on the side.

FIGURE 1.18 Classroom connecting terraces.
Source: Render by Hartz (2020); adapted from: Uribe (2016).

The heating comprised floors with radiant panels, allowing the entire space to have a uniform temperature without causing annoying air and dust circulations, even when the windows were open. An inspection report carried out on July 27, 1948 recognized that the school satisfactorily met the required conditions, from the pedagogical, hygienic, and physiological point of view. The building currently retains its complete form, even with the passage of time and the evolution of education (Uribe, 2016) (Figure 1.19).

FIGURE 1.19 View of the building in Amsterdam.
Source: Render by Hartz (2020); adapted from: Uribe (2016).

The goal at that time was to create healing buildings, and even clinical white colors were part of the design, dust was avoided in decorative parts.

1.3.6 COVID-19

The most recent pandemic occurred at the end of 2019 in the city of Wuhan, China, identified as a coronavirus (COVID-19), on March 11th, 2020, the WHO recognized it as a global pandemic. The population of the entire world was asked to quarantine, stay inside their homes, and leave as little as possible. All public spaces were closed to avoid the transmission of the virus, but this evidenced that most houses are not designed for this.

Using the words of Leilani Farha of the UN, "*housing has become the first-line defense against the coronavirus*," who shows that, in the world, there are around 1.8 billion people who do not have adequate housing or are homeless (Figures 1.20 and 1.21).

Also, Leilani Farha points out that inadequate housing refers to one that lacks quality materials, suffers from overcrowding, or does not have basic services (Urban Center, 2020), aspects that determine life's quality, but over time these parameters have lost importance, relegating them even by the same users not only from the authorities, however, COVID-19 made visible the problem which makes us rethink, what are the conditions to live in a healthy place?

FIGURE 1.20 A house made of urban waste housing in a vulnerable area in Torreón, Coahuila, 2015.
Source: Photograph of the author.

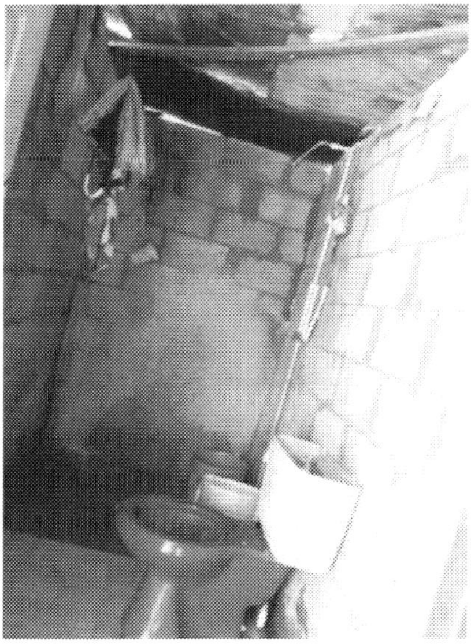

FIGURE 1.21 Bathrooms in a vulnerable neighborhood, 2015.
Source: Photograph of the author.

Nowadays, there are many aspects we take as normal, without realizing that the evolution and changes of the furniture, spaces, openings, symbols, regulations, and standards, have been and will be determined by needs related to health. In general, infectious diseases are those generated by microorganisms, such as viruses, bacteria, fungi, and parasites, which can be transmitted through direct contact with infected patients, their blood, or secretions (Unión de Mutuas, 2019), and hygiene is required, coinciding with the requirements of past pandemics.

Alternative measures against the spread of pandemics are the closure of schools, theaters, or the prohibition of public events, to help the mitigation, which together with the early implementation of confinement measures imposed by governments, delay peaks infections (Chaparro, 2020).

But the question is, what is the challenge that architecture must face in the face of this pandemic? Some actions have been put on:

- First, it is a healthy distance, which breaks the paradigm of social distancing, which is related to the capacity to occupy spaces, where signs are required to establish the distances to be considered between users (Figure 1.22).

FIGURE 1.22 Social distancing in the municipal office of Ramos Arizpe, 2020.
Source: Photograph of the author.

- Second, consider the time of the active virus on the surfaces by providing protection for the spread of the virus between users when distancing is not possible (Figures 1.23 and 1.24).

FIGURE 1.23 Examples of protections in work, 2020.
Source: Photographs of the author.

FIGURE 1.24 Examples of protections in toilet areas in public or service areas, 2020.
Source: Photographs of the author.

- Third, ventilation, and air renewal in closed spaces help to dilute the virus in the area, even more so if a certain number of people are concentrated.
- Fourth, control access of entrances and exits to check people's temperature with enough space between users, and if it is possible, the entrance should be far from the exit, although sometimes buildings do not have the appropriate dimensions (Figures 1.25 and 1.26).

FIGURE 1.25 Strategies to control the entry of the virus through the users, 2020.
Source: Photographs of the author.

FIGURE 1.26 Strategies to control the entry of the virus through the users, 2020.
Source: Photographs of the author.

- Fifth, and not least the access to hygiene services; however, there are still places where the supply is not good or is scarce.

There are protocols that can be a guidance to spatial design; nevertheless, these recommendations are omitted. Something important to highlight is that these pandemics have similar requirements and over time have been forgotten (Table 1.1).

TABLE 1.1 Comparison of Design Strategies Generated by Pandemics

Period	Pandemic	Opportunity Areas	Alternatives
XIV Century	Plague	• Lack of hygiene	• Distancing • There were no documented proposals in spaces design
1489	Typhus	• Overcrowding • Absence of light • Bad ventilation • Dirt	• Distancing • Hygiene • Natural light • Ventilation • Cleaning • There were no specific proposals
XIX Century	Cholera	• Lack of hygiene • Bad ventilation • Lack of sewage • Overcrowding	• Sanitary drainage • Ventilation in bathrooms • Adequate spaces for the number of people • Bathrooms • Air renewal in spaces • Spaces with natural light
From the 7th century until now	Measles	• Overcrowding • Cleaning • Indifference to get vaccinated	• Distancing • Cleaning • Hygiene • There are no proposals, only consider easy-to-clean surfaces
Modernist period (1918)	Influenza	• Overcrowding • Dark and dusty spaces	• Minimalist concept where the decoration on the surfaces were zero • Easy to clean surfaces • Ventilation for air renewal • Light colors, mainly white • Interior spaces connected to exteriors to promote a healthy space • Entry of light to sanitize the spaces • Prevent the entry of dust

TABLE 1.1 *(Continued)*

Period	Pandemic	Opportunity Areas	Alternatives
2020	COVID-19	• Crowding • Constant lack of hygiene • There is no distancing between people • Lack of clean surfaces	• Social distancing • Avoid being in crowded spaces • Ventilation that promotes air renewal in interior spaces • Control area when accessing a space and, if it is possible, direct access to a bathroom • An area for shoe cleaning (special installation to place the sanitizing covers) • Ample space to keep distance, especially in public areas. • Smooth surfaces for easy cleaning or coating that does not allow the virus to remain or reproduce. • Propose flexible spaces that can be expanded or reduced • Consider expanding dimensions to have a distance of 1.5 to 2 meters between users

Source: Own elaboration.

The letter of Athens of 1933 and what it says in Part II, is still valid. In *Current State of Cities. Critiques and Remedies*, the room observations in point 9 indicate:

The admissible density for buildings is 250 to 300 inhabitants per hectare. When 600, 800, and even 1,000 inhabitants are reached, then they are slums, characterized by the following signs:

- Insufficient living space per person;
- Mediocrity of openings to the outside;
- Lack of sun (orientation to the north or consequences of the shadow that falls on the street or in the patio);
- A permanent presence of morbid germs (tuberculosis);
- Absence or insufficiency of sanitary facilities;
- Promiscuousness in the interior layout of a home, a poorly organized property, or the presence of annoying neighborhoods.

Over the centuries, vegetation has been replaced by hard surfaces and the green surfaces, which are the lungs of the city, have been destroyed. With these conditions, permanent discomfort and illness occur.

Point 10 says:

> *In congested urban sectors, living conditions are dire due to a lack of sufficient space for accommodation, lack of available green areas and, finally, lack of maintenance of buildings (due to speculation). Which is aggravated by the inability of people to adopt defensive measures by themselves.*

This is reflected by the narrowness of the shady streets and the total lack of green spaces, to improve air quality, and a healthy place for children to recreate.

Point 12 says:

> *The constructions intended for living beings are distributed by the surface of the city, contrary to hygienic needs.*

The first duty of urban planning is to adapt to the fundamental needs of men. The health of each person depends, in large part, on their submission to natural conditions, the sun, which presides over the entire growth process, should penetrate the interior of each home to spread its rays, without them, life withers; air, whose quality ensures the presence of vegetation, should be pure, free of harmful gases and dust suspended in it. Finally, the space would have to be widely distributed. It should not be forgotten that the sensation of space is psychophysiological, and that the narrowness of the streets or the narrowing of the avenues create an atmosphere that is as unhealthy for the body as it is depressing for the spirit. At the same point are indicated requirements that should be met for all, but this has not yet been achieved.

1.4 FACTORS THAT MODIFIED ARCHITECTURE AND URBAN PLANNING

Not only pandemics modify architectural environments, but there are also other artificial or man-made elements, which makes us rethink everything we know.

1.4.1 MAN AND HIS AMBITION

During the first half of the 19th century, at the height of the Industrial Revolution, the installation of industries and demographic expansion collapsed the cities surrounded by walls. The old Spanish urban centers were not only the cradle of the processes of densification and urban speculation but also the origin of the first complaints about the extreme living conditions that affected the weakest social groups. In Spain, as in other parts of Europe, the first voices of alarm about the living conditions in the cities were by doctors and hygienists. In 1841, in Barcelona, Pedro Felipe Monlau, denounced the hygiene conditions (Laboratori D'urbanisme, 1992). Commercial profit, the high control by the government and other factors have maintained until today these conditions of overcrowding and lack of access to basic services, which allow the proliferation and lack of control in the health aspect.

1.4.2 DESTRUCTION BY FIRE

In 1871, Chicago was attacked by a fire that practically destroyed it, and multiple factors that changed the way of thinking about architecture influenced the reconstruction of the city. In its first decades the city was built of wood using the Balloon Frame technique, since it could be built quickly and with no specialized labor (Benévolo, 2002).

The catastrophic event generated a change in the vision of architects and engineers at the time of the reconstruction, breaking the traditional paradigm of building and designing a city. This movement was known as the Chicago school and the trend was to build vertically, using new techniques like a steel skeleton and another type of foundation, giving the opportunity to create the first electric elevator and the telephone, having hotels and warehouses with many floors, including the use of extended large windows, and the result was an esthetic balance between vertical and horizontal lines (Benévolo, 2002).

In 1895 someone who visited Chicago expressed:

> *The construction of office buildings of enormous height, with an iron and steel skeleton structure that supports the internal and external walls, has become a custom in almost all the great American cities. This construction style was born in Chicago, at least in its practical application, and this city now has more buildings of the steel skeleton type than all other American cities combined (cited by Benévolo, 2002).*

1.5 EFFECTS ON THE ENVIRONMENT

1.5.1 ATMOSPHERIC POLLUTION

Clean air is one of the basic requirements of health and well-being. However, the development of industrial and post-industrial societies has considerably increased the presence of polluting particles in the atmosphere, which is damaging to human health and the environment. Air pollution is any change caused by an unnatural external agent that alters the physical and chemical properties of the air, mainly derived from industrial processes that involve combustion, in industries, cars, and heating systems (Soler and Palau, 2018, p. 3).

The General Law of Ecological Balance and Environmental Protection defines the environment as:

> *"The set of natural and artificial elements or those induced by man that make possible the existence and development of human beings and other living organisms that interact in a space through time" (Article 3, Section I).*

Humans have a close relationship with the environment and nature, so the quality of life, health, and even the present and future material and cultural heritage are linked to the biosphere. In other words, the life of human beings depends on the life of the planet, its resources, and species (CNDH, 2016). WHO estimates that over 90% of the population live in places where air quality is not taken care of, but not all pollutants come from outside, inside buildings there are activities that modify air quality, due to use of cleaning chemicals, sprays, and materials used in the building's construction.

Air pollution is one of the main causes that deteriorate the quality of life in cities due to the harmful effects it has on people's health (Soler and Palau, 2018, pp. 13) and the lack of green spaces that minimize this impact.

1.5.2 AIR QUALITY

We will start from two laws:

- In the General Law of Ecological Balance and Environmental Protection (2018), the environment is defined as the set of natural and artificial elements or those induced by man that make possible the existence and development of human beings and other living organisms that interact in a determined space.

- In the General Law of Climate Change (2018), environment is the variation in climate attributed directly or indirectly to human activity, which alters the global atmosphere composition and adds to the natural variability of the climate observed during comparable periods.

Both laws indicate how humans adapt their habitat for their own benefit, but this change has gradually affected the planet, and is now being reversed against the inhabitants. This can be observed as how the climate in different areas changes over the years, resulting in the spaces inside buildings no longer responding in the same way as they did before, cities gradually increase their temperature as there is no balance between surfaces artificial and natural.

The problem is the exponential growth of the population without adequate planning, the exploitation of natural resources, the alteration of ecosystems, and the contamination of water, soils, and air; among others (CNDH, 2016).

We classify sources of contamination of human origin into four major groups (CNDH, 2016; WHO, 2018):

1. **Transportation:** Motor vehicles, airplanes, trains, ships, and the consequent handling of the fuels they use.
2. **Stationary Combustion:** Homes, businesses, industrial energy, including thermoelectric plants.
3. **Industrial Processes:** Chemical, metallurgical, refineries, paper mills, etc.
4. **Others:** Agricultural burns, garbage dumps, fires, leaks, spills.

1.5.2.1 CASE STUDY

In Coahuila, as in Mexico City and others, there is concern about air quality, which is seen in the increase of discomfort and diseases associated with pollution. Authorities have quantitatively evaluated its quality, spatially, and temporally, and were supported by national and international standards that determined if the air quality was satisfactory or not. Therefore, the Secretary of Environment of the State of Coahuila (SMA) has implemented an air quality monitoring program in the city, and they have three fixed automatic monitoring stations.

Where the following standards are considered:

- NOM-025-SSA1-2014. DOF, Official Mexican NORMA NOM-025-SSA1-2014, Environmental health. Permissible limit values for the concentration of suspended particles PM10 and PM2.5 in ambient air and criteria for their evaluation.

- NOM-020-SSA1-2014. DOF, Official Mexican STANDARD NOM-020-SSA1-2014. Permissible limit value for the concentration of ozone (O_3) in ambient air and criteria for its evaluation.
- NOM-021-SSA1-1993. DOF, Official Mexican NOM-021-SSA1-1993, Environmental Health. Criterion for evaluating ambient air quality with respect to carbon monoxide (CO).
- NOM-022-SSA1-2010. DOF, 2010. Official Mexican STANDARD NOM-022-SSA1-2010, Environmental health. Criterion for evaluating the quality of ambient air with respect to sulfur dioxide (SO_2).
- NOM-023-SSA1-1993. DOF, Official Mexican NOM-023-SSA1-1993, Environmental health. Criterion for evaluating the quality of ambient air with respect to nitrogen dioxide (NO_2).

They consider the regulated pollutants for which a maximum concentration limit has been established (Table 1.2).

TABLE 1.2 Pollutants and Their Nomenclature

Pollutants	Chemical Formula
Ozone	O_3
Carbon monoxide	CO
Nitric oxide	NO
Nitrogen dioxide	NO_2
Nitrogen oxide	NOx
Sulfur dioxide	SO_2
Suspended particles smaller than 10 microns	PM10
Suspended particles smaller than 2.5 microns	PM2.5

Source: Own elaboration based on SMA.

In addition, they measure meteorological parameters to model pollutant emissions:

- Temperature;
- Humidity;
- Atmospheric pressure;
- Direction of the wind;
- Wind speed;

- Solar radiation;
- Precipitation;
- UV radiation.

The results of the air quality monitoring are evaluated under the scale Metropolitan Air Quality Index (IMECAS) (Table 1.3).

TABLE 1.3 Flagging based on the IMECA Scale

Flag	IMECA Points	Air Quality
	0–50	Good
	51–100	Regular
	101–150	Bad
	151–200	Very bad
	201	Extremely bad

Source: Own elaboration based on SMA (2020).

The impact of air pollution directly affects health according to the WHO (2018) (Table 1.4).

TABLE 1.4 Pollutants and Their Impact

Pollutant	Health Effects
Sulfur dioxide	Headache, anxiety, and eye and throat irritation
Suspended particles smaller than 10 microns	Impact on the respiratory system
Suspended particles smaller than 2.5 microns	Cardiovascular and nervous system diseases
Ozone	Respiratory diseases
Nitrogen dioxide	Breathing problems and asthma
Carbon monoxide	Cardiovascular disease and neurobehavioral effects

Source: Own elaboration based on SMA (2020) and WHO (2018).

The continued exposure to air polluted with high levels of solid particles reduces life expectancy, being an important aspect to consider in any space, whether open or closed.

In 2007, the Saltillo Secretary of the Environment monitored seven areas of the city with a high traffic density to determine the air quality, resulting in that the least critical is Boulevard Morelos and the most critical reaching the Unsatisfactory level with suspended particles corresponded to Boulevard Isidro López Zertuche and Periférico Luis Echeverría Álvarez due to the great anthropic impact. These avenues lack green areas, in addition to having nearby the company TUPY, Vitromex, many shops and gas stations. Two years later, an increase in pollutants can be seen in other areas, although the most critical continue to be Boulevard Isidro López Zertuche and Periférico Luis Echeverria.

The latest data recorded by SMA on air quality was in 2017. Until October, the SMA showed that the highest percentage of pollutants occurred in January and May with PM10; even though the most constant was PM2.5, generating poor air quality, followed by ozone. Most affect the respiratory, nervous, and cardiovascular systems. The problem is that PM particles include pollen, microscopic biological material, dust, recirculating soil, soot, and other small solids.

But it should be noted that pollution not only comes from the outside, it can also be generated inside the same construction, so it is important to always take into account good ventilation, since we can have two invisible and sometimes silent enemies.

1.5.3 INDOOR AIR QUALITY

In a construction, the walls play an important role in achieving a healthy environment; the materials used together with ventilation and natural light are relevant, if the space is hermetic to save energy and does not allow good ventilation, it generates poor air quality and does not allow sanitation, proliferating microbiological contamination that causes allergies, asthma, among other diseases, due to the prolonged stay in an unhealthy space damaging health, sometimes air quality is not compatible with energy efficiency. Ashrae points out that "*the main obligation is to achieve healthy and comfortable buildings and that energy saving is in the background*" (cited by Higuero, 2016).

1.5.3.1 SMELLS

This is another polluting element that may not be so harmful, but it does generate discomfort outside and inside buildings, generated by different sources, such as factories, garbage, among others, caused mainly by man by not considering environmental smells when constructing a building.

The Technical Measurement Standard in Baubiologie MSB (2015) comments that:

> *Smells are first perceived with the nose. Intensity and quality play the leading role. In addition, it is possible to distinguish if the smell is pleasant or unpleasant, or if there are concrete indications of the cause (fungi or chemical products) (pp. 34).*

For this, it is necessary to search for alternatives that help to mitigate them, in case it is not possible, the source or the capture point can be moved considering the direction of the wind, for example, placing a screen with trees that captures or disguises odors or channeling the unwanted winds through natural or artificial elements, but the first alternative will always be to study the place in detail to detect sources of contamination before building.

1.5.3.2 THE RADON

Radon (^{222}Rn) is a naturally occurring radioactive gas from the decomposition chain of ^{238}Uranium and is therefore ubiquitous in nature. It can accumulate inside buildings, and its disintegration products can be inhaled and deposited in the bronchopulmonary tract. Radon is currently considered the main source of natural radiation exposure for humans (CTE, 2019, p. 144).

WHO (2016) indicates:

> *Radon leaks through cracks in floors or at the junction of the floor with the walls, spaces around pipes or cables, small pores in walls built with hollow concrete blocks, or through sinks and drains. The concentration of radon in a home depends on the amount of uranium contained in the rocks and the soil, and it is mainly concentrated in living spaces that are in direct contact with the ground, such as basements and cellars.*

Depending on the type of construction, the ventilation habits of its inhabitants vary from day to day and the tightness of the building. Therefore, when constructing a building, one must consider the exposure to radon, especially in geological areas with a high concentration of this gas.

In the Technical Building Code (CTE, 2019) in Section HS6 in point 2 it states that:

> *To limit the risk of exposure of users to inappropriate concentrations of radon from the ground inside habitable premises, a reference level is*

established for the annual average concentration of radon of 300 Bq/m³ (pp. 138).

It is important to know the terrain, but also to ensure that the spaces have adequate ventilation to have a constant renewal of air that does not allow the air in the space to be contaminated.

Section 3.1 protection barrier, point 1, indicates the importance of having this element that limits the passage of gases from the ground, having the following characteristics of point 2 in the same section:

- To have continuity in joints and sealed joints;
- To seal passageways or similar;
- The communication doors that interrupt the continuity of the barrier must be watertight and equipped with an automatic closing mechanism;
- Avoid cracks that allow the passage of radon from the ground by convection;
- To have adequate durability for the useful life of the building, its conditions, and the planned maintenance (pp. 139).

The Technical Measurement Standard in Baubiologie SBM (2015) indicates that the radon concentration in a building fluctuates strongly over time, being important that together with the ventilation of the indoor air, consider the outdoor climate and the fluctuations in temperature and pressure, as well as the characteristics of the soil.

- In winter, because of calefaction, Rn concentrations are noticeably higher because of higher temperatures, poorer ventilation, and concentrations of air from the ground.
- In summer, radon concentrations in indoor spaces are usually up to five times lower than in winter. Also, in the subsoil, there can be clear differences in radon gas concentration due to the seasons, but the differences are usually smaller and are approximately 1.5 to 3 times.

Problems in the indoor atmosphere can also occur from open, nuclide-rich (pp 14) building materials.

The WHO (2015, p. 60) indicates that it is important that the strategy must be applied not only to new homes, but also to existing ones, mitigation should not be the only objective, but also prevention to reduce the concentration of this gas in the homes and even in buildings where there is a high concentration of people who have a long stay inside the building. The structural systems, foundations, and ventilation should be studied, considering that the conditions and propagation mechanisms may vary from one place to another, either by infiltration of gases through the ground, emanation of construction materials and through water.

Radon control options in a construction can be (USEPA, 1993, p. 61):

Option	**Tracing**	**Functioning**
Active ventilation of the sanitary chamber	Periodic following of radon	Moderate
Active soil depressurization (it is the most common)	Periodic following of pressure and radon	Moderate
Balanced ventilation (between the air flow extracted from a space and the one blown inside)	Periodic following for radon	Moderate to high

Source: Own elaboration based on USEPA (1993). They establish other options, but the rest were low to non-existent.

Ventilation of non-living spaces between the ground and habitable space such as vented sanitary chambers, which can help reduce indoor radon concentrations by separating the interior from the ground and reducing radon concentration below the habitable space.

For it to work, it depends on several factors, such as the degree of air tightness of the floor located above the non-habitable ventilated space and, in the case of passive ventilation, the distribution of the ventilation openings along the perimeter of the non-habitable space, a variant involves the use of a fan to pressurize or depressurize the non-habitable space, but it can generate problems such as reverse draft in combustion appliances or energy losses (ASTM, 2003 cited by WHO, 2015).

It must be considered that ventilation presents uneven results, and can cause energy losses, especially in extreme climates. But if the main source of radon is building materials, ventilation is necessary (WHO, 2015).

1.5.3.3 CO_2

Carbon dioxide (CO_2) is one of the most common pollutants found inside buildings and affects human health. CO_2 is generated by household appliances (Serrano, 2017) and by humans when breathing, although it is harmless.

The ranges to be considered regarding CO_2 concentration according to Serrano (2017) and Siber (2016) are:

- Typical outdoor CO_2 concentrations: 350–450 ppm (parts per million).
- The normal range in a CO_2 home should be between 400 to 600–800 ppm.
- Tolerable CO_2 concentrations in IAQ: 1,000 ppm.
- The annual average concentration must be less than 900 ppm in each location.

There must be a minimum flow of 1.5 L/s per habitable premises in non-occupancy periods, with respect to CO_2 generation values, it is considered 19 L/h per occupant, or 12 L/h per occupant for the period sleep and 19 L/h per occupant for the waking period.

The number of occupants to be considered for calculating the CO_2 generated will basically depend on the number of bedrooms:

Amount of Space	Number of Occupants
1 bedroom	2
2 bedrooms	3
3 bedrooms or more	4

Source: Own elaboration based on Serrano (2017).

In the main bedroom, two occupants are considered. In the others only one occupant, in the bathrooms 0.5 L/h per occupant is considered. For this reason, all passive and mechanical systems must be capable of maintaining recommended levels of ventilation according to the activity, occupation, and location with respect to the climate (Serrano, 2017). A person resting takes about 12 breaths per minute, which involves mobilizing about 360 L/h.

At night, a high amount of CO_2 can be concentrated if it has little ventilation with one or two people, and in the morning the stale air can be perceived, for this it is advisable to open the windows at least five minutes so that the space is ventilated which allows to restore values of 500 to 600 ppm, improving air quality (Villena, 2018). Since a high level of CO_2 in the environment can cause effects on health or performance (Serrano, 2017).

1.5.3.4 HUMIDITY

Humidity plays an important role not only in health but also in the construction itself. To control humidity, ventilation in the space is important, along with other measures, the high percentage of humidity in a space prevents sweat evaporation of the human body, making it impossible to eliminate body heat (Siber, 2016), when the air is saturated with humidity, it cannot absorb it, generating discomfort in space. It is important to remember that when temperature drops, humidity increases, favoring the growth of microorganisms like mold, on the contrary, if a high temperature is maintained, other organisms such as *Legionella pneumophila* proliferate, so the selection of materials used in a construction is crucial, so those do not serve as substrates for these microorganisms. There are standards that establish the parameters and considerations for health spaces and equipment, such as ISO 14644-1, the CTN 171 – indoor environmental quality, UNE 171330, UNE 171330 standards-2: 2014 Indoor environmental quality, Part 2: Interior environmental quality inspection procedures, and the UNE 100012 standard whose objective is to define a method to assess the hygiene of air conditioning systems, among others.

The Department of Health and Human Services (2005) notes:

> *People exposed to a damp, moldy environment can have several health effects, or they may not have any problems. Some people are sensitive to mold. For these people, mold can cause nasal congestion, throat irritation, coughing or wheezing, eye irritation, or, in some cases, skin irritation. Mold is found indoors and outdoors and grows in places with high humidity.*

Moisture causes problems for building owners, maintenance personnel and occupants. The American Society of Heating, Refrigerating, and Air Conditioning Engineers (ASHRAE, 2016) notes that, in many cases, common humidity problems can be due to poor decisions in design, construction or maintenance.

An environment with a high humidity rate negatively affects the well-being and mood of the occupants because of bad smells and the perceived poor air quality (Siber, 2016). Humidity control is important for the well-being of the occupants and for the conservation of the building, furniture, and facilities, for the following reasons (Table 1.5).

TABLE 1.5 Relationship of the Percentage and its Effects

Humidity Percentage	Effect
If the relative humidity is below 30%	Wooden furniture tends to dry out and crack
Under 40%	Viruses, bacteria, and respiratory disorders appear
If the relative humidity exceeds 40%	Produces oxidation on ferrous metals
If the relative humidity is below 55%	Generates static electricity in synthetic carpets and furniture
Above 60%	Produces fungus, mold, and allergy in people
For people who do not have chronic problems	Humidity should not exceed 65%
For people with health conditions	Humidity must not exceed 60%
If the relative humidity is high, more than 90%	People can be uncomfortably hot during the summer, because the evaporative cooling of the body through sweat is suppressed
If the relative humidity in winter is not too low	People feel more comfortable and do not suffer from dry, cracked skin
If the relative humidity is high in winter	There will be condensation on cold surfaces
Between 40 and 60% relative humidity	It is the ideal level in a closed space

Source: Own elaboration (2020); Siber (2016).

The different regulations on energy efficiency and the implementation of mechanical ventilation systems generate an improvement in air quality and reduce gas emissions; in the case of allergy victims, it helps a lot to improve life's quality of, but they are still insufficient for healthy environment (Soler and Palau, 2018, p. 14). There are materials that, through photocatalysis, i.e., to the exposure of natural or artificial light to a certain specific wavelength, are activated and eliminate pollutants, or others that, through zeolites in plasterboard, purify the air (Maroto, 2016).

Finally, the Technical Measurement Standard in Baubiologie MSB (2015) specifies that it should be considered to carry out inspections in buildings and to the users themselves through visual and olfactory inspections; if relying on safety data sheets, techniques, construction certificates, and photographic documentation, it is recommended:

- To inspect the interior spaces;
- To consult residents about the history of the building;

- To check the materials used, equipment, furniture, flooring, adhesives, paints, lacquers or other construction and renovation materials;
- To investigate current or past episodes of odors, suspicions, or symptoms of illnesses.

The visual inspection must be exhaustive, including information related to the construction of the floors, walls, and roofs, use of auxiliary premises and attached houses, habits of use and aeration, which helps to find weak points in the construction according to the year of construction (pp. 24).

1.5.3.5 FUNGI

The technical measurement standard in Baubiologie SBM (2015) on fungi states that:

Indoors should not have visible signs of mold or fungi, nor contamination by spores or their metabolites:

- The number of mold fungi in the indoor atmosphere, on surfaces, in dust, in gaps, in materials, etc., should be lower than the exterior or at the same level as the unaffected comparison rooms. The type of mold inside it is not always the same from that on the outside or in the unaffected comparison rooms.
- Particularly critical and toxin-producing fungi that are allergenic, or that thrive at a body temperature of 37°C, should not be detectable at all or only extraordinarily little.
- Long-term high humidity in materials and air, as well as cold surface temperatures, should be avoided as they are the basis for fungal growth.

Significant details, suspicions or indications of microbial contamination must be detected, for example, discolorations, and stains, characteristic odors of fungi that indicate humidity, construction damage, even more so in constructions with hygiene problems, with a humidity contribution from the outside above of the average, with pathologies of the past, the history of the building, diseases of the users, etc. In some cases, fungi appearance is just the tip of the iceberg of microbiological contamination, fungi are a base element for recycling organic waste that returns assimilable substances to the environment. As a result of humidity and hygiene problems, bacteria very often appear together with fungi, which generates problems for the health of the inhabitants (pp. 37).

These microorganisms not only affect surfaces or furniture, but also inhabitants' health, 10 to 20 critical agents have been studied in interior spaces such as endotoxin, bacteria, or viruses. Fungi proliferate due to the conditions of temperature, humidity, and light, in a warm humid climate the conditions are favored, instead with constant ventilation and dry air it disseminates.

Some considerations would be the use of materials with good hygroscopicity, since it reduces the accumulation of moisture on the surface, it is important that the materials are dry before use to avoid the accumulation of moisture. Capillarity in the enclosures should be avoided to reduce the conditions for mold proliferation, reduce thermal bridges and improve indoor ventilation conditions. Air conditioning equipment must be maintained in optimal conditions (Table 1.6).

TABLE 1.6 Source and Type of Microorganisms in the Spaces

Source (if is not well done from the beginning)	Microorganism
Painting on surfaces, ceiling, wallpaper on surfaces or in a plaster	They generate fungi
Human beings	Bacteria, fungi, and viruses
Surfaces where there is no constant cleaning or that are not easy to clean	Bacteria, fungi, and viruses
Mechanical systems that are not given periodic maintenance and that do not renew the air	Bacteria, endotoxins, fungi, viruses, and amoebae
Composting area, if it is not located far from the spaces and is in the direction of the winds	Not only pollutes with bad odors, it generates fungi that can enter the interior of a house (a minimum distance of 10 meters is recommended)

Source: Own elaboration based on Figols (2016).

1.6 COVID-19 BREAKING THE PARADIGM OF ARCHITECTURE 2020

In the 21st century, the SAR-COVID-19 pandemic is not different from the previous ones, and it is making changes again. It makes us rethink the paradigm of spaces and social distancing, in addition to hygiene. It should be considered that the normality before this pandemic is not going to return, and we need to prepare ourselves. The virus is here to stay along with other viruses and bacteria, and it has come to shake and reveal what has been hidden under the carpet.

The pandemic does not consider economical, race or social aspects, it could have started with people with high resources or constant international mobility for work or study, but upon reaching the countries it was distributed without distinction, although the most affected population due to confinement was the most vulnerable, poor people with limited resources, without a permanent job, forcing them to not comply with the quarantine, some even do not have a roof.

The emergence of the pandemic exposed the housing crisis in recent years and exacerbated it exponentially. The connection between the reproduction of the virus with the type of urbanization is very visible; the higher the concentration, the greater the spread. Overcrowding multiplies the spread of the virus; the population that was invisible, became visible due to the saturation of the health system. The question is, did a pandemic have to come to understand that housing and the development of a dignified life is a human right? (Di Filippo, 2020).

The answer is no, and in the academia and in architecture research and related areas, the housing crisis had already been exposed for a long time, but the decision-makers have not yet been able to do anything, even with the flag of sustainable social housing, this is not fulfilled in some cities of Mexico (Figures 1.27–1.29).

FIGURE 1.27 An example of a sustainable social housing, 2017.
Source: Photograph of the author.

FIGURE 1.28 The access to sustainable social housing, 2017.
Source: Photograph of the author.

FIGURE 1.29 The access to sustainable social housing, 2017.
Source: Photograph of the author.

An example of this is presented by Ibarra and López (2019) who remark that in Guadalajara, private companies are always reducing costs in the production of social interest houses that directly affects the location on cheap and poorly located land, with a lack of facilities and infrastructure or in the worst case, located in vulnerable areas. Using cheap and fast materials in their construction process, which are not suitable for the climate, and with spaces with minimal measures based on the regulations with the sole objective to build three houses where two or one is better; in situations of a prolonged stay, it generates stress and other physical and emotional pathologies, due to discomfort, and resulting in a damage society.

When talking about space, it has been observed that closed social housing areas in Mexico have been reduced to a minimum mobility to such a degree that only the bed and a small space to store clothes fit, in addition, the limited space affects the capacity for privacy. The kitchen is a basic service area and, in most cases, is ridiculously small, which makes it impossible to carry out activities with more than two people. Another aspect is the mobility for the elderly, disabled or temporarily disabled, the most critical area being the bathroom, which has been punished in its dimensions and prevents adequate access to its interior, resulting in a reduced home or mini housing due to its insufficient, deficient, and even uninhabitable dimensions for human beings with reduced mobility to carry out daily tasks.

Since May 2020, a discussion was opened about what was happening and what will come, from government and academic levels, highlighting several aspects, for example, in the webinar entitled *When housing fails to be a refuge*, organized by the Universidad de los Andes, held on May 12, master's students of architecture, lawyers, and anthropologists from Colombia highlighted the impact of staying at home:

- Lack of water service;
- Good ventilation and lighting inside the houses;
- The systematic reduction of spaces in homes, including those of social interest;
- Lack of network connectivity to be able to work from home;
- Commuting from work to home;
- Evictions;
- Overcrowding.

In another webinar titled *Housing after COVID-19* webinar (UCLG, 2020); including people from different governmental levels and internationalities from London, Barcelona, France, Canada, and UN-Habitat; held on May 22, it was showed that:

- Resilient cities are those that help social cohesion;

- That housing must be affordable;
- Social distance is a problem for spaces with a high concentration of people;
- Community housing should be commissioned by non-governmental groups;
- Houses should meet quality standards;
- Home offices are on the rise and it is possible that it is a trend which saves rent money;
- Those who are homeless.

In the webinar *Conditions of homes habitability and the urban environment in the face of social isolation imposed by COVID-19*, held on May 22, coordinated by UNAM, the first report carried out in eight Mexican was shown, focusing on 10 axes: on typology, services, activities carried out at home, use of open and closed public space, family, and neighborhood coexistence, security, and perception of government action, highlighting similarities and differences in each city:

- The greatest activity that increased at home was domestic;
- Half of the people in the study indicated that the government's actions are correct;
- The majority said that the house is their own and is paid for;
- The problem of paying for housing and services;
- Space for the elderly;
- Home-work connection;
- Lack of equipment in their neighborhood;
- Lack of accessibility to public spaces, mainly those with low resources.

Even when they are from different environments, they coincide on some problems and show the opportunity areas that this pandemic has brought to light. UN-HABITAT, points out seven indicators of adequate housing and COVID-19, which focus on:

1. **Secure Tenure:** Housing is not adequate if its occupants do not have a certain measure of security of tenure that guarantees them legal protection against forced eviction, harassment, and other threats.
2. **Affordability:** Housing is not adequate if its cost endangers or hinders the enjoyment of other human rights by its occupants.
3. **Habitability:** The home is not adequate if it does not guarantee physical security or does not provide sufficient space, as well as

protection against cold, humidity, heat, rain, wind or other health risks or structural dangers.

4. **Access to Basic Services, Materials, Facilities, and Infrastructure:** Housing is not adequate if its occupants do not have drinking water, adequate sanitary facilities, energy for cooking, heating, lighting, and food preservation, or waste disposal.
5. **Accessibility:** Housing is not adequate if the specific needs of disadvantaged and marginalized groups are not considered.
6. **Location:** Housing is not suitable if it does not offer access to employment opportunities, health services, schools, daycare centers and other social services and facilities, or if it is in contaminated or dangerous areas.
7. **Cultural Adequacy:** Housing is not adequate if it does not consider and respect the expression of cultural identity.

Since the quarantine was directed in Mexico, many researchers have carried out the activity of documenting information about the confinement and its consequences, the Academic group of Technology in Architecture of the Faculty of Architecture in the Saltillo Unit, of the Autonomous University of Coahuila, together with The Ministry of Housing and Territorial Planning in contact with ONU-HABITAT, developed an instrument to measure the perception of people regarding their environment and the way of inhabiting the house in this stage of confinement called THE PARADIGM OF COVID-19 IN ARCHITECTURE in the Southeast Metropolitan Area of Coahuila, reaching 600 surveys, exceeding the representative sample (384 surveys), with data from other cities outside the state, in total 462 corresponded to the studied area.

The survey was online with 29 items, considering some UN HABITAT indicators with response options using the Likert scale, to people aged over 15 years, disseminated through social networks and emails, previously a pilot test was carried out at the end of April and changes were made in three items, mainly in writing and response options, and the survey was distributed throughout the month of May 2020. This chapter will only show some of the results related to the subject of the book, the objective is to monitor during the rest of 2020 and 2021, to observe the differences throughout the quarantine, the adaptation to the new reality and after of the confinement, considering all the UN-HABITAT indicators and the location of the respondents to obtain a map and compare it with contagion areas identified by another government office to establish direct policy actions in the state of Coahuila.

Figure 1.30 shows that 81.1% of the population has their own home, only 13.4% rent a property, and the rest lived in a borrowed house or did not respond. In a second stage, the survey included if they had payment problems while paying or renting it, the majority responded that they finished paying it, in June, the Mexican government proposed alternatives together with banks to support debtors, but the issue of income is another aspect that is still under discussion.

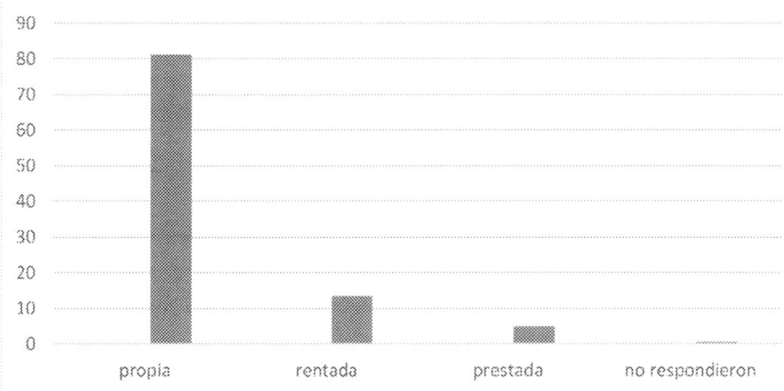

FIGURE 1.30 State of the properties of the respondents: Owned, rented, or borrowed.

Figure 1.31 shows that 33% of the families comprised four members, 55% had three rooms, the percentage of more than six members with less than three rooms was low; 19% of the homes were overcrowded, coupled with this was the type of users.

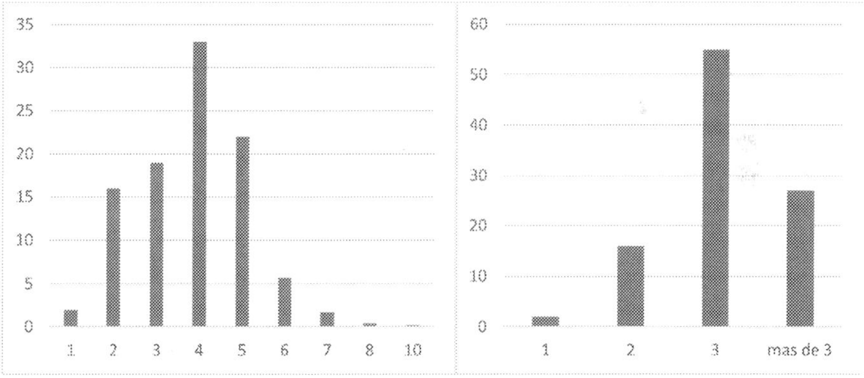

FIGURE 1.31 Overcrowding was analyzed regarding the number of inhabitants (a); and the number of rooms (b).

Figure 1.32 shows that 50% are adults or young people, 27% live with children, 18% with the elderly, and a low percentage with people with disabilities (PWD) or have with these three types of inhabitants, being an important aspect to consider in terms of coexistence and space limitations based on the type of construction.

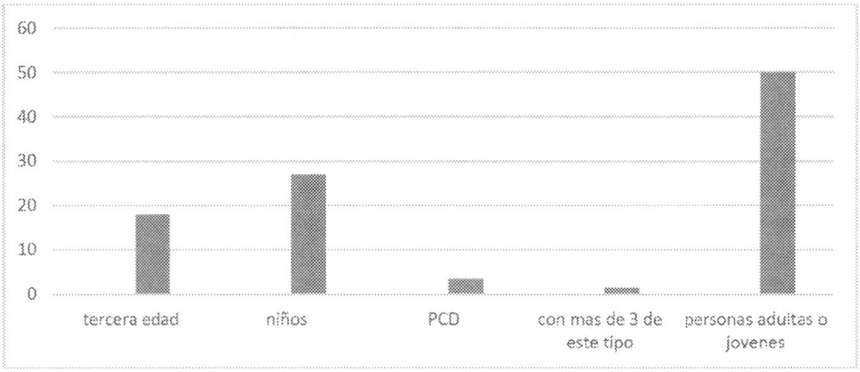

FIGURE 1.32 Characteristics of the inhabitants.

Figure 1.33 shows that 92% are families, 3.9% with pets, and the rest of the relationships are friends, alone, or work colleagues, so family coexistence predominates in this metropolitan area of Saltillo, although in other cities may not be the case anymore, due to migration for college or work.

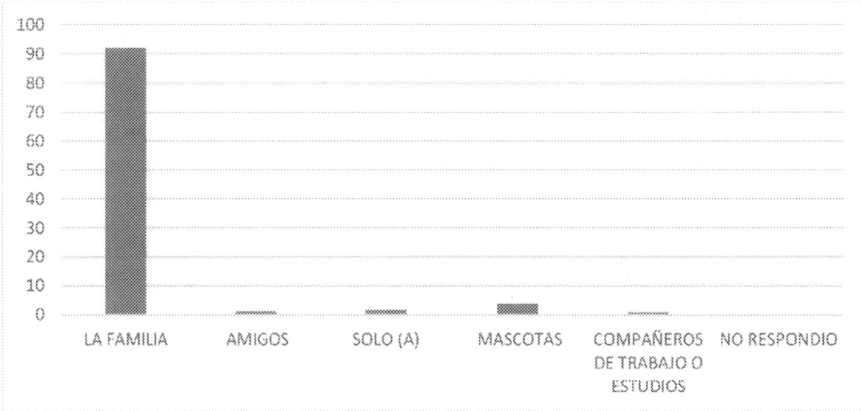

FIGURE 1.33 Relationship between the inhabitants.

Figure 1.34 shows that 66% live in a two-floor construction, 31% on a one-floor house, which is more convenient when it has elderly, children, and PWD living in it, the rest live in an apartment, the number of people is also important when considering hygiene and health.

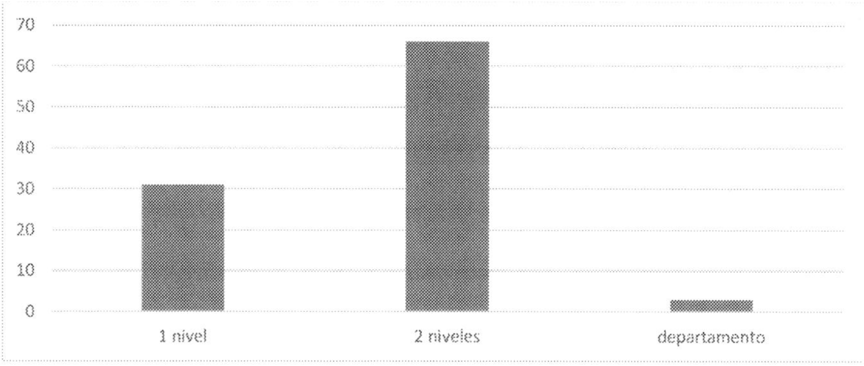

FIGURE 1.34 Construction typology of the houses in the study.

Figure 1.35 shows that 41.3% indicate that people have three or more bathrooms in their home, 22% only have one bathroom. If the number of people living in a house is not greater than three, it does not generate so much conflict, but in the case of the COVID-19 or another pandemic where hygiene control is required, this could be considered in order to control the sources of infection among those who live in the house, requiring alternatives to minimize the risk. Therefore, flexibility in the home should be considered when adapting to new needs.

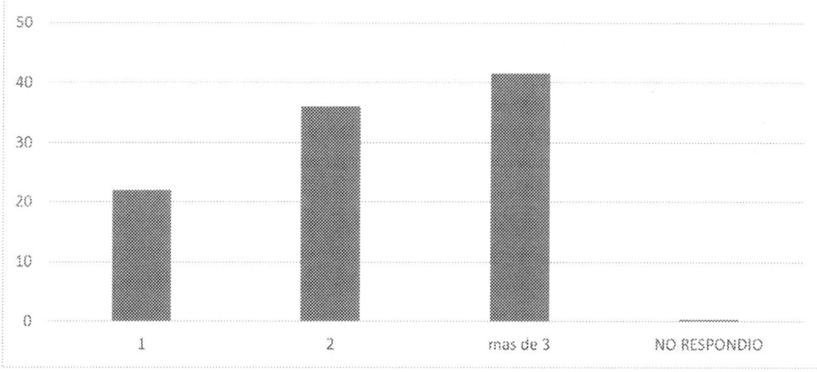

FIGURE 1.35 Number of bathrooms.

Figure 1.36 shows that 27% consider that their home is flexible to adapt new spaces, 31.6% say that it is possible, and 15% do not believe that their home can be adapted, due to spatial limitations.

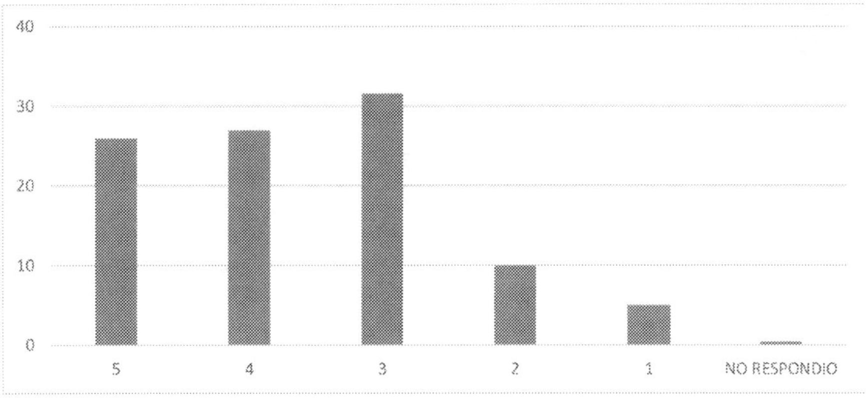

FIGURE 1.36 Flexibility of the space inside the house.

Figure 1.37 shows that 43% think their home is comfortable, 32% say that it is more or less, and 9% say that their home is not comfortable. This does not only include the thermal sensation inside the house, but the perception of tranquility, and that allows you to carry out activities inside the house.

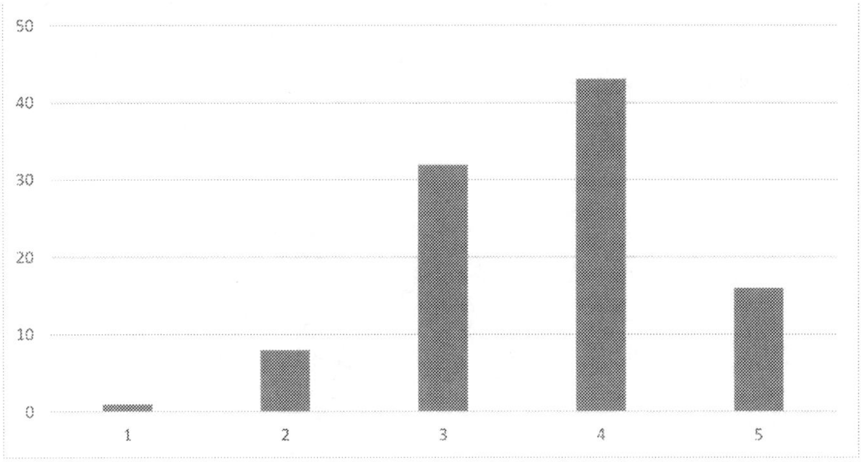

FIGURE 1.37 Comfort in the houses.

At this point, it is important to specify that studies carried out by Kampf, Dodt, Pfaender, and Steinmann (2020), indicate that the virus at temperatures of 30 to 40°C with a humidity of 30% reduces its duration on a surface, at room temperature the virus can last up to nine days.

Additionally, it has been demonstrated that temperature and ventilation reduce or increases the virus contagion, for example, in open spaces there is a lower contagion possibility, as well as in closed spaces with ample ventilation (Instituto Nacional de Bolivia cited by FM Bolivia, 2020). The key is good ventilation, because recent studies have showed that virus particles are so small, that can remain suspended in the air for some time, so it is relevant to keep the space aerated and not use a mechanical system to recycle it, as it generates a source of infection among the people who use that space.

In Figure 1.38, 44.2% feel calm inside their home, only 4.4% do not feel calm in their home. This is an opportunity area, to work in a space that can help people feel good in their homes, more so now that they must stay for longer. For this, it is important that the house can adapt to other activities that meet specific needs at certain times where it must be contained for more than 8 hours and have a peaceful space to rest. An important aspect is to keep noise pollution low or null to reduce stress for the inhabitants or prevent them from working or studying, in this long period of confinement.

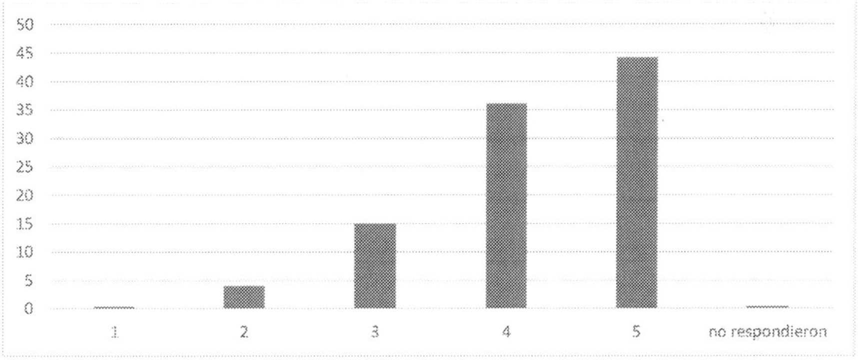

FIGURE 1.38 Perception of tranquility.

In Figure 1.39, 54% indicated that their home did not require drastic changes to develop other activities that were not common before, 19% adapted a space to take classes, other activities that predominated during this pandemic were doing sports followed by family activities. In a low percentage was work from home, businesses, and even dancing lessons,

0.2% adapted a containment space in the access to their home, a special area to control the entry of the virus to the house.

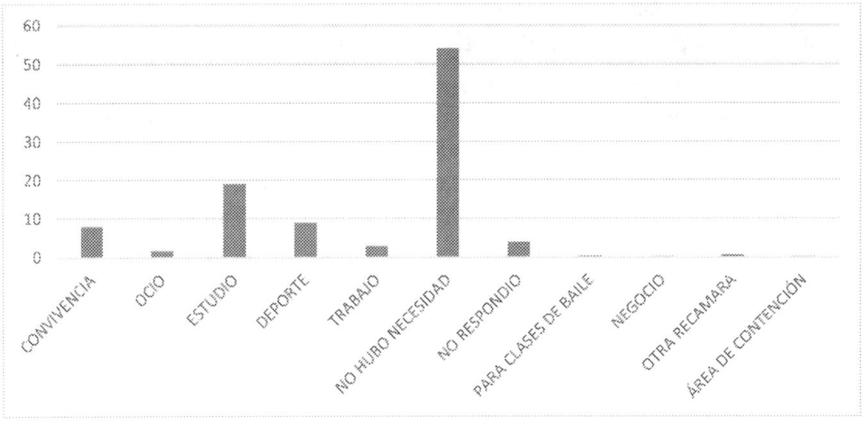

FIGURE 1.39 Adequacy of the space for other activities.

Figure 1.40 shows that 53.2% of the people fully agreed to have this space; when considering the role it plays in the health of the inhabitants of the home, only 2.2% did not agree, 10.2% considered taking it into account.

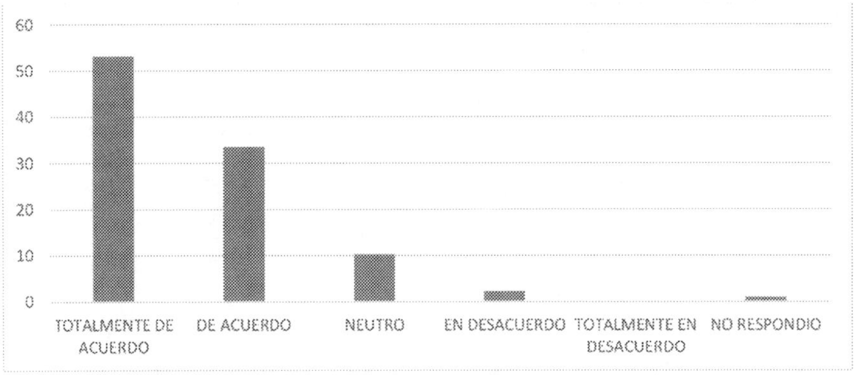

FIGURE 1.40 Containment space or access filter.

In particular, this space has been established with safety regulations in areas with a high concentration of people such as work and public spaces, but it should also be taken into account for homes, especially when there is one or two people who leave constantly, in addition to using accessories such as a sanitizing mat for footwear, having a space where to put street shoes and place footwear commonly used for the house, with a shelf to have a sanitizing gel, which allows the person coming from outside to be sanitized and reduce the introduction of the virus into the home as little as possible.

In some cases, there are houses that have a hallway, garage, or bathroom where you can sanitize upon arrival. Besides, it is recommended that the person when arriving home maintain a healthy distance from those who are inside, and if possible, stay isolated, in case they work in a place of medium and high risk, along with other actions such as bathing right away and wash their clothes separately from others.

Figure 1.41 shows that 36.6% had not realized if they kept a healthy distance when they got home, 29% claimed to have a distance of 1.5 m, and 12% indicated that they had a distance less than 1 m. These are interesting data and give an answer as to why the increase in infections in the population, people have confidence and do not maintain a safe distance to avoid infecting others. Where social distance is difficult to maintain, hence importance of considering to include a space that serves as containment and control before accessing a closed space.

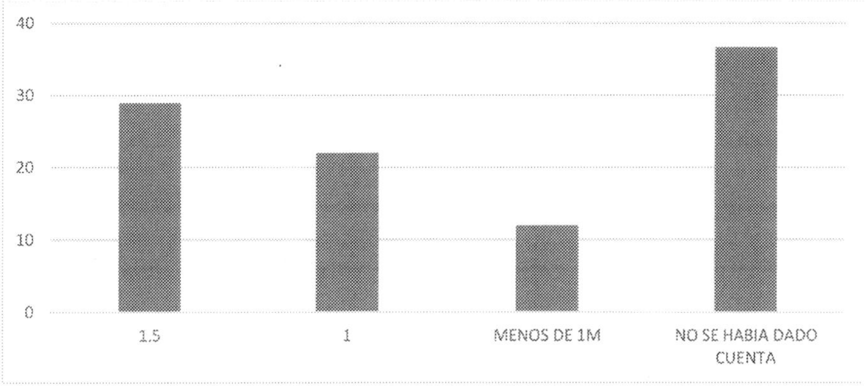

FIGURE 1.41 Healthy distance when arriving home.

In Figure 1.42, when asking the question, if you agree to change some surfaces in your home to make it easier to clean viruses and bacteria, 57.3% agreed, 6.7% did not agree, and 17% had not even thought about it. Evidence from some scientific tests shows how the virus persists for longer on some surfaces than on others, here is the relevance of rethinking what materials to use in buildings to help maintain a healthy home.

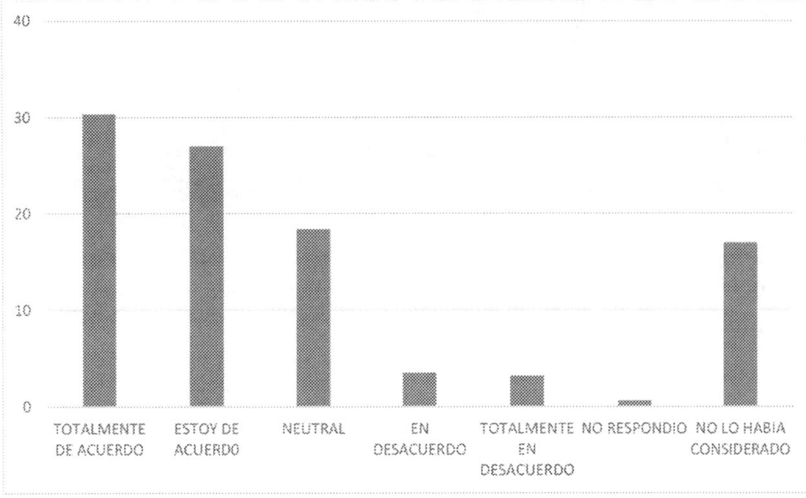

FIGURE 1.42 Surfaces in the home.

It is important to consider the time the virus stays on surfaces and materials, even more so if someone is or was sick, studies carried out by Kampf, Dodt, Pfaender, and Steinmann (2020), indicate that the virus can last from 2 hours to 9 days on a surface (Table 1.7) there are other studies from the Paraguayan Society of Infectology and the University of Hong Kong, which in some data coincide and in others differ in the days or hours, in addition, they indicate that the temperature also plays a role.

TABLE 1.7 Persistence of SARS-COV of Different Types of Inanimate Surfaces

Type of Surface	Temperature	Persistence
Metal	Room temperature	5 days
Wood	Room temperature	4 days
Paper	Room temperature	4 to 5 days
Glass	Room temperature	4 days
Plastic	22–25°C or room temperature	4 to 9 days
Disposable gown	Room temperature	2 days
Aluminum	21°C	2 to 8 days
Surgical glove (latex)	21°C	8 hours
Steel	4–40°C	48 hours to 28 days
Cardboard box	there is no data	24 hours

Source: Own elaboration. Based on Kampf, Dodt, Pfaender y Steinmann and Sociedad Paraguaya of Infectología quoted by Ultimahora (2020)

The importance of knowing the duration of the virus on surfaces is that cleaning is carried out easily and at least twice a day on these, mainly in common areas, avoiding porous surfaces that are not easy to sanitize.

Another aspect to consider from being at home is having basic services, such as electricity and water, but now another service has become the main actor for work, study, and leisure at home in this quarantine and it was the internet.

Figure 1.43 shows that 41.1% had no network service problems, 10.4% indicated that the quality of the network was not good during the pandemic. This service previously considered a luxury, now plays an important role not only for leisure, but also for work, study, and commerce. In the case of remote work and study, a good connection and easy internet access should be considered, designating a different space and hours, if there are several who are going to use it (CMIC, 2020), but this is sometimes not easy when the house is small and another matter to consider is the noise pollution from the outside, since the buildings are not designed to generate an environment that allows these activities to be carried out properly, nor does the environment help to achieve these conditions.

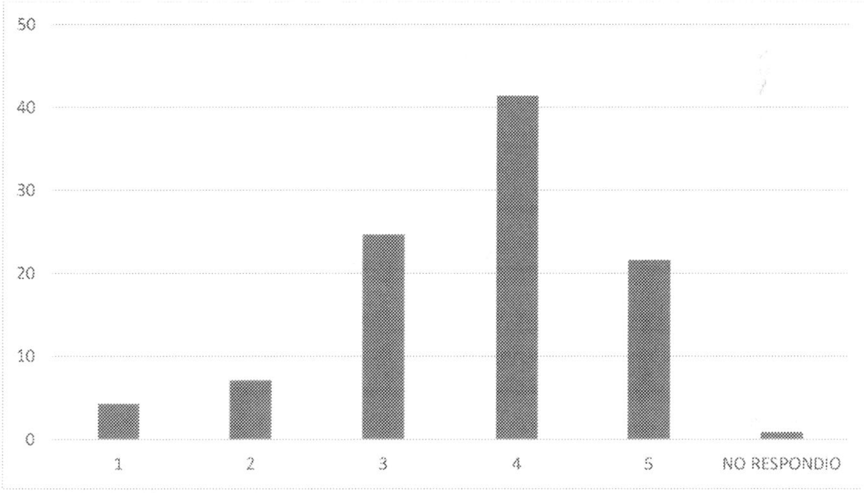

FIGURE 1.43 Internet services in the houses.

Figure 1.44 shows that 78% indicated that they had a very good water supply, and only 5.2% indicated that the service was not adequate or was being supplied constantly. This natural element is essential for life, and crucial for health control through constant hygiene, but it is scarce in some parts of the world, it becomes the most requested protagonist in this quarantine. That is why the importance of guaranteeing the supply of this liquid everywhere.

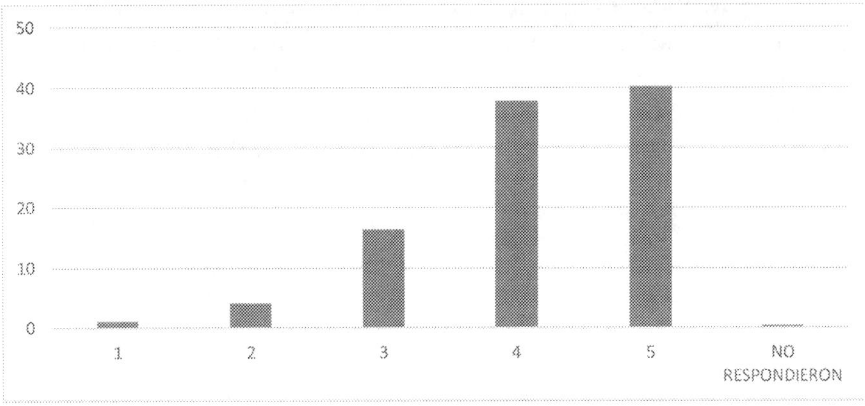

FIGURE 1.44 Water services.

In informal settlements there is a vulnerable niche in the face of the COVID-19 pandemic. And it is that the lack of services, such as water and drainage, poor quality materials in a house, but the families that inhabit it in a risk of contagion (Centro Urbano, 2020).

Figure 1.45 shows that 48.1% have a mall near their home, 21% said they have a shopping center nearby and the rest have access to public transportation, which are basic. There are three scenarios of the pandemic (IMSS, 2020) in relation to social distance; when there are dozens of cases, no action is necessary in open or closed public spaces, but when there are hundreds or thousands of incidents, activities, and events must be suspended.

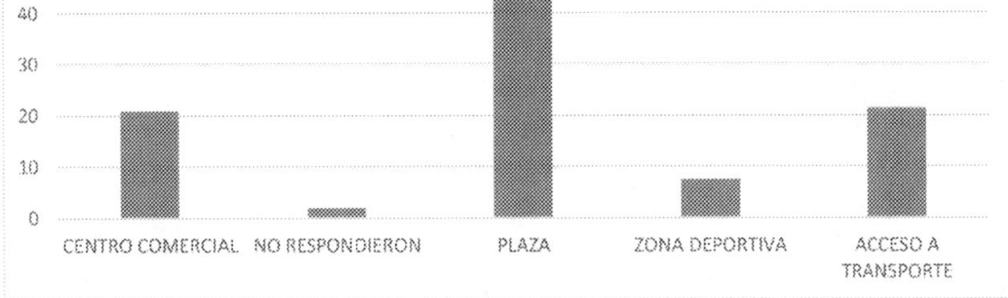

FIGURE 1.45 Another indicator are the stores close to the home.

When phase 3 of the pandemic is reached, which consists of the reopening activities, open public spaces are the most demanded by the population to practice sports, meeting with people and recreation, this is considered medium risk (Government of Mexico, 2020).

With the new normality or reality, other parameters are established to consider in the different scenarios of a city (Table 1.8) based on the amount of agglomeration and the type of ventilation in the site (Table 1.9). These criteria are met in most cases, but sometimes it is relaxed by both the employees and the users themselves, even in some companies they control and monitor their staff to reduce contagion and use strategies such as rotation of schedules to lower the number of people in the spaces.

TABLE 1.8 Considerations and Priority Level

Closed Public Spaces	Workspaces	Business Spaces	Open Public Spaces	Housing Spaces
Very high priority	High priority	Medium to high priority	Medium priority	Low priority
Entrance and exit	Entrance and exit	Entrance and exit	No control	Entrance and exit
Cinemas	Government offices	Offices with cubicles	Plaza	Houses
Stores	Industries	Production companies	Parks	Public roads
Museums	Banks	Beauty and hairdressing salons	Private vehicles	
Hospitals	Elevators	Malls	–	–
Public transport	Cinemas	Restaurants	–	–
Family reunions	Supermarkets	Doctor's office	–	–
Bars and discotheques	Study centers (preschool to graduate)	Supermarkets	–	–
Religious events	–	Refuge	–	–
Massive events	–	Shelters	–	–
Gyms	–	Drugstores	–	–

Source: Own elaboration; National Institute of Bolivia and Colombia (2020) cited by FM. Bolivia and Microsoft News.

TABLE 1.9 Considerations to Adapt or Design Spaces

General Considerations for Enclosed Spaces
To access the buildings, you must have control where the entrance is, which should be separated from the exit, the temperature should be taken with a laser thermometer or a thermal camera, a sanitizing mat together with the gel to sanitize the hands when entering, throwing away the garbage that from outside and garbage from the inside in special containers, even more so when they wear disposable masks, in some cases they have bathrooms where staff can shower before leaving, depending on the activity they carry out.
Spaces at 50% of their capacity or less according to the phase, but if the pandemic phase is red or orange, the percentage may be lower.
Surfaces that are easy to clean.
Maximum capacity of people within a space, in special cases have a control for appointments so as not to have excess capacity.
Constant ventilation, in case of using a mechanical system that the air renewal does not recycle the air.
Distances of at least 2 meters between workspaces and have separating elements that help maintain isolation between people in case someone does not use a mask.

Source: Own elaboration.

Public places will have to be adapted to avoid physical contact and mitigate contagion, avoid contact with surfaces that carry viruses and bacteria. It should be considered that the doors are automated, as well as the elevators, and implement an automated temperature control. Logistics are required to apply these adaptations to the new reality, as well as the resilience to make it work. In places of constant movement of tourists, adapting private spaces or few people, as well as in meeting places such as lobbies or elevators, limiting the passage or contact of several people at the same time should be considered, but taking care of the details to not having a cold architecture that isolates everyone is a task that should be considered and avoided.

The Mexican Institute of Social Security (2020) points out that after this pandemic, the way of acting with the environment should change and be more empathetic with the most vulnerable, privileging the right to a healthier life. Another variable must also be added, which was proposed before the pandemic and climate change; pointed out by Lionel Ohayon, founder of the New York design and urban planning study ICRAVE, he comments that telecommuting and the redesign of work cubicles had to be rethought by other more closed and personal ones, with technology and reducing social contact. If virtual work is successful or generates more production, the value of shared workspaces will change, considering the creation of spaces in homes designed for work or studies with the conditions that allow developing this new activity without distractions.

1.7 QUALITY STANDARDS FOR COMFORT, WELL-BEING, AND HABITABILITY

Previously, architects were sought to create a place to live, at the end of the First World War, due to the destruction of cities, mainly in Europe, the accelerated growth of constructions, including houses, was promoted, initiating massive construction, breaking with the patterns above, prioritizing rational distribution. In Mexico, this construction process was based on institutional programs that over time lost the objective of adequate housing for just a commercial product, resulting in the reduction of spaces and materials that are cheap, but not suitable for all types. In addition, the particular needs of its inhabitants are not considered, having a direct impact on the quality of life of users.

Sometimes the bricklayer or the engineer or even the user himself creates his designs or takes the idea of an architect, and based on this he builds his home, resulting in most cases, spaces that fail to cover the basic needs that meet the standards of a healthy and comfortable home, since it is not only creating a space, it is taking into account other elements.

Since its inception, architecture has laid its foundations on three Vitruvian pillars: constructive solidity, functionality, and beauty. However, the experience of space, whether in natural, urban, or indoor environments, is also an emotional experience, so that another fundamental aspect underlies the user's emotional response, closely related to well-being (Guixeres, Higuera, and Montañana, 2016), which generate experiences (Figures 1.46–1.48).

1.7.1 WELLNESS

The role of the architect when constructing a building or house is to make the user's dream come true, since it is not only a physical space, it is the habitat that contains the cultural, social, and psychological aspects, generating sensations or covering emotional needs. The question is, what experiences do we generate in the spaces and if this is positive or negative?

FIGURE 1.46 The experience in interior spaces in Morocco, 2005.
Source: Photographs of the author.

Shapes together with color and light play an important role when designing spaces, since they generate various sensations in the user such as joy, sadness, phobia, or stress, which affect not only physical but also psychological health, which is related to well-being, the WHO (1945) defines health as a state of complete physical, mental, and social well-being, and not only the absence of affections or diseases.

FIGURE 1.47 The experience in exterior in Romania, 2007.
Source: Photographs of the author.

FIGURE 1.48 The experience in exterior in Vienna, 2008.
Source: Photographs of the author.

Designing implies considering all the architectural elements that provide security, ease of mobility, adequate visibility, which contribute to the well-being of the user and the pleasure of feeling good most of the time during their stay in the space where they live, and this is related to the comfort.

1.7.2 ENVIRONMENTAL QUALITY

This concept applies to the interior of buildings not only covering air quality, but also health, comfort, esthetic, anthropometric, ergonomic, acoustic, and light conditions and considering the existence of electromagnetic fields in the site or its immediate surroundings.

1.7.2.1 HEALTH AND COMFORT

UN HABITAT (2010) comments:

> *Everyone has the right to an adequate standard of living, including adequate housing. Despite this, the number of people without adequate housing far exceeds 1 billion worldwide, living in dangerous conditions for life or health, overcrowded, and makeshift settlements (pp. 1).*

The relationship between health and comfort of the people cannot be ignored in the buildings where they live, work, study or have fun (Siber, 2017), in other words, where we live most of the time.

Reckford (2009), points out that:

> *The quality standards are about a living environment rather than a house, and they establish that the function of a house must be fulfilled for the family that inhabits it, that is, to provide protection against the weather, an appropriate living space, adaptations of a house that are culturally acceptable and access to adequate public services of drinking water and sanitation, transportation, and socioeconomical benefits (pp. 4).*

Comfort is subjective, it depends on many factors and aspects, in general it describes the perception and feeling of well-being of the subject in a given place and time that allows them to carry out an activity without the distraction of discomfort or being thermally, physically, and mentally uncomfortable in which are involved:

- Age;

- Complexion (metabolism);
- Culture (place of origin, adaptation to its environment);
- Race (skin color);
- Health;
- Activity.

It can be evaluated qualitatively and quantitatively, the latter based on well-defined, measurable, and observable parameters using devices. The qualitative refers to the perception of the subject that is a physical stimulus to his surroundings through the five senses (smell, taste, touch, sight, and hearing).

The role of each of the senses actively participates with architecture, the texture of the materials awakens interest through sight, generating the need to touch the surface to corroborate what one thinks will feel, the smells that focus on the space can bring to mind childhood memories or tell us if we are close to something or transport us to a specific space, for example, smelling popcorn reminds us of the cinema since that space has that identity, so through this classic smell, taste also participates, finally, sound is a crucial element in a construction, if it is handled properly you can get total isolation from the outside if you do not take into account the morphological aspects of the nearby environment, surfaces, and the shape of the building can have characteristics that absorb sound, reflect it and even direct it to a particular point, the knowledgeable architect and sensitized about this, takes advantage of the architectural elements to obtain something unique.

Quantitative comfort encompasses the thermal, acoustic, and visual aspect, which when harmoniously combined result in physical and psychological well-being, whether in an open or closed space. Most of the time, humans are inside a closed space, therefore, it is important to achieve a healthy space. The human habitat must have conditions to avoid a high percentage of mechanical systems, these should only be complementary or for support, the temperature and air quality depend on achieving comfort and health.

Bueno (1998) points out that a construction should not be hermetic, since it is not ensured that it breathes, recommends having a balance between energy savings and air quality, that there must be a relationship with the climate and the geographical relationship of the construction based on selecting bioclimatic strategies and passive systems that solve specific problems according to the bioclimatic zone, in accordance with the ASHRAE decalogue, which indicates prioritizing comfort and health before energy saving.

1.7.2.2 VENTILATION

This parameter is relevant when designing spaces, since it is part of the analysis of the terrain, the first thing is to identify the surroundings of where it is intended to build, if there are visual pollutants and odors, which could affect the level of comfort and health of the users. To establish a strategy that reduces the negative impact, through modifications in the morphology of the subdivision, analyzing the wind direction and implementing the vegetation as an ally, insulator or as a protective screen, the placement of buildings to redirect the wind in favor or opposing.

It is important to know not only the direction of the wind and its speed for air renewal, but also the quality of the air and unwanted noise, to consider natural elements that help reduce this pollution before entering the interior spaces of the constructions, there are currently subdivisions that are located near municipal garbage dumps, factories, schools, main roads or very close to these, which indirectly affect the well-being of users.

Knowing the behavior of the fluid, it helps to get an idea of how it moves through the existing elements in the environment (Figure 1.49) and how it enters the interior of the spaces, besides considering the Venturi effect, which is key to increase or decrease the wind speed, for the benefit of the inhabitants.

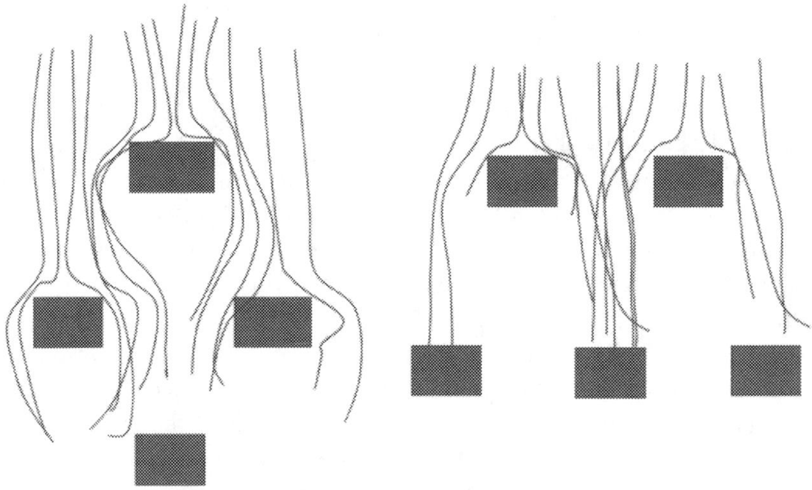

FIGURE 1.49 Wind movement through buildings.
Source: Own elaboration based on Fuentes and García (2005).

Air is a fluid that always seeks to follow its movement, all the elements disturb the wind regime laminar as turbulent, having an idea of how the movement is between the buildings, gives the possibility of placing them, that they do not remain in the shadow of the wind and make it impossible to take advantage of the favorable winds to renew the air indoors. On the other hand, if there are undesirable winds such as bad smells or cold winds in winter season, the shade can be used in favor to avoid them, this being the best strategy. If you have a dense barrier and a high permeability barrier, the wind speed varies by 20 to 25%. The influence of this reduction comprises 200 meters.

Even when it is not possible to have an opening in the orientation of the dominant or secondary winds, it is possible to use constructive or natural elements that channel the wind to the desired space (Figure 1.50).

Even when it is not possible to have an opening in the orientation of the dominant or secondary winds, it is possible to use constructive or natural elements that channel the wind to the desired space (Figure 1.50).

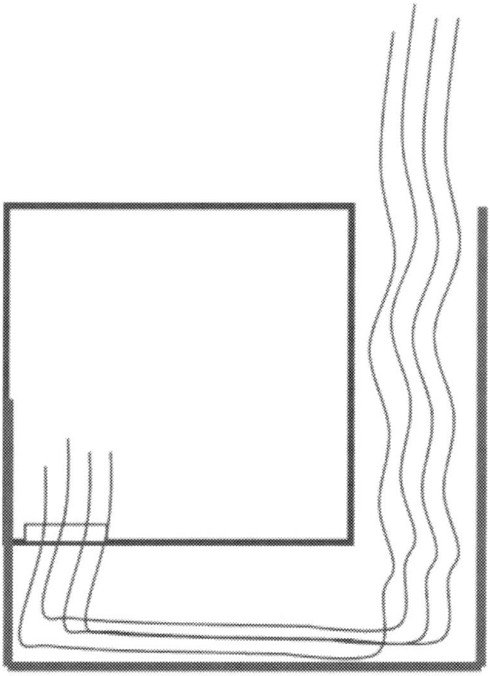

FIGURE 1.50 Redirection of the wind by means of another element.
Source: Own elaboration.

Fluids always look for an exit, opposite openings generate air mobility inside the constructions (Figures 1.51 and 1.52), but this mobility is due both to the speed it has when entering and the difference in temperature with respect to the spaces, if there are obstacles inside, the air will seek the shortest path, if the surfaces are rough, the speed of the fluid will be reduced.

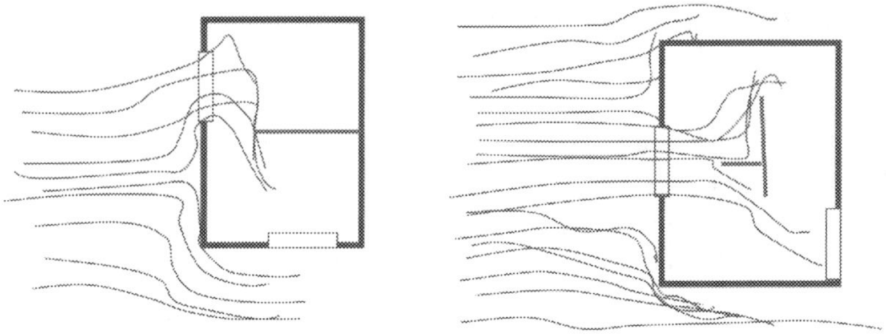

FIGURE 1.51 Movement inside the construction in a floor.
Source: Own elaboration based on Fuentes and García (2005).

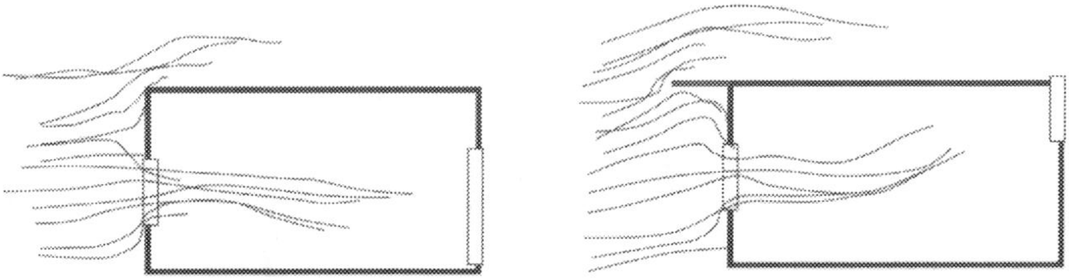

FIGURE 1.52 Movement inside an elevated construction.
Source: Own elaboration based on Fuentes and García (2005).

It should be remembered that the objective of good ventilation is to supply enough fresh air, to extract emissions from the activities that are carried out by the occupants and their appliances, together with the humidity and dangerous substances that are inside the building. To dehumidify the space, the ventilation rate must be high.

1.7.2.3 MECHANICAL VENTILATION

The use of mechanical systems in interiors plays an important role, especially when climatic conditions or activity demand environmental control, in microbiological proliferation, for which several aspects must be considered (Table 1.10).

TABLE 1.10 Relationship of Air and Pollutants

Source	Result	Recommendation
Poor outdoor air quality	Proliferation of microorganisms	Test the air quality of the environment
Filters	Fungi	Constant cleaning of the filters and renew them every so often
Stagnant water in humidifiers	Biological agents such as bacteria	Constant cleaning
Air recycling	Contaminates existing biological agents	The extracted air goes directly to the outside

Source: Own elaboration based on Figols (2016).

It is important to always have a control of humidity and temperature, along with an adequate air renewal based on the activity and capacity of users.

1.7.2.4 AIR RENEWAL

Air renewal has a role to achieve comfort in the spaces by providing fresh air, and it must be completed with a system that allows the evaluation of the polluting products that have accumulated in the stale air mass, because a person resting absorbs, approximately about 27 liters of oxygen per hour under normal conditions and expels 23 liters of CO_2. The optimal level of air movement indoors should be between the limits of 1.50 m/s during the day and 1.00 m/s at night (Molar, 2014; Siber, 2017; WHO, 2018).

The main drawback of natural ventilation is regulation, since the renewal of each moment will depend on the weather conditions, the size of the openings and the orientation it faces outside.

Permanent natural ventilation depends on fixed values. Molar (2014) says:

- The urban morphology of the immediate surroundings, which help channel the wind or divert it, increase its speed, or reduce it;

- Constructive characteristics of the building;
- Natural elements near the openings;
- Artificial elements of the same building that direct or obstruct the access of the wind to the interior;
- Orientation and position of the building;
- Position and dimension of the openings, these must guarantee the number of renovations necessary to ensure optimal indoor air quality, depending on their shape and type of window;
- The contour of the building must be as tight as possible, to avoid unwanted ventilation;
- The building can play the role of air reservoir, ensuring the quality of the outside air without constant ventilation.

The Technical Measurement Standard in Baubiologie SBM (2015) comments that:

> *The renewal of the air in a building depends on many factors such as the tightness of the building, the exterior and interior climate, the season of the year, the wind and pressure circumstances in the environment and within the building, the position and dimension of windows, open aeration over windows, as well as technical ventilation installations (decentralized or centralized) using fans (pp. 35).*

The Technical Building Code (CTE) in Spain (2019), establishes the minimum flows of closed spaces (m^3) (Table 1.11), since it indicates that all homes must have a ventilation system, either hybrid or mechanical, recommends that the air must circulate from the dry to the humid places, the kitchens must have an additional specific mechanical system independent of the ventilation duct of the rest of the house.

TABLE 1.11 Minimum Flow Rates for Constant Flow Ventilation in Habitable Premises

Housing Type (m^3)	Minimum Flow q_v in l/s				
	Dry Places (1) (2)			Humid Places (2)	
	Main Bedroom	Bedrooms	Living Rooms and Dining Rooms (3)	Minimum in Total	Minimum per Location
0 or 1 bedrooms	8	–	6	12	6
2 bedrooms	8	4	8	24	7
3 or more bedrooms	8	4	10	33	8

TABLE 1.11 *(Continued)*

Housing Type (m^3)	Minimum Flow q_v in l/s				
	Dry Places [(1) (2)]			Humid Places [(2)]	
	Main Bedroom	Bedrooms	Living Rooms and Dining Rooms [(3)]	Minimum in Total	Minimum per Location
In the cooking zone of kitchens (considered as a humid place)	A system must be in place to extract the pollutants produced with a minimum flow rate of 50 l/s. This space must have good air extraction to generate constant renewal.				
Bathrooms	It is a humid place				

[(1)] In the dry premises of the dwellings for various uses, the flow corresponding to the use for which the highest flow results is considered.

[(2)] When both dry and wet premises are used in the same room, each area must be provided with its corresponding flow.

[(3)] Other premises belonging to the dwelling with similar uses (gaming rooms, offices, etc.).
Own elaboration based on the Official State Gazette (2017); CTE (2019); and Siber (2020).

The air renewal rates based on the technical measurement standard in Baubiologie SBM (2015) for other rooms (Table 1.12).

TABLE 1.12 General Renovation

Space	Air Renovation (m^3/h)
Open office	40–50
Individual office	40
Classroom, auditorium, and inn	30–40
Conference room	30
Theater, concert, and cinema	20
The minimum hygienic renovation	0.3

Source: Own elaboration; Technical Measurement Standard in Baubiologie SBM (2015), p. 35.

In the Basic Document (DB) in section HS, of article 13 of point 1, it states:

> *The objective of the basic requirement "Hygiene, health, and protection of the environment," is to reduce to acceptable limits pollutants that risk users inside buildings and under normal conditions, when suffer discomfort or illness, as well as the risk on deteriorated buildings which deteriorate the environment in their immediate surroundings, as a consequence of the characteristics of their project, construction, use, and maintenance (pp. 3).*

In Section 13.3 on the basic requirement HS 3: Indoor air quality, from point 1:

> *Buildings should have enough resources so that their enclosures can be adequately ventilated, eliminating the pollutants that occur regularly during the normal use of the buildings, so that a sufficient flow of outside air is provided and the extraction and expulsion of the air stale by pollutants is guaranteed (pp. 3).*

In Section 3.2.6 exterior windows and doors, from point 1:

> *The exterior windows and doors that are available for supplementary natural ventilation must be in contact with a space that has the same characteristics as that are required for intake openings (pp. 70).*

In Section H3 regarding air quality, in general in point 2:

> *In the habitable premises of the dwellings, a sufficient outdoor air flow must be provided to ensure that in each premises, the average annual concentration of CO_2 is less than 900 ppm, and that the annual accumulated CO_2 that exceeds 1,600 ppm is less than 500,000 ppm per h (pp 63).*

If CO_2 concentration values in the outdoor air are not available at the building site, an annual average of 400 ppm is considered, Appendix C, point 3 (pp. 80).

The big problem with traditional heating and air conditioning is that they do not create a circulation that extracts poor quality air, nor do they allow humidity or bad smells to escape. If you have the windows closed to keep warm in winter and cool in summer, a stale air mass is generated that can cause disease, discomfort, and bad odors, among other negative effects (Siber, 2016), generated by enemies that persist in the area such as humidity and other contaminants that are not perceived, such as mites and germs that exist in closed spaces.

1.7.2.5 LIGHTING AND RADIATION

A properly lit space generates tranquility and harmony, the opposite result in a dark setting that gives the feeling of insanity and insecurity. The combination of light with color makes a space appear large or small, happy, or sad, being light distribution better with light colors (Figures 1.53 and 1.54).

FIGURE 1.53 Natural light and space, 2005.
Source: Photographs of the author.

FIGURE 1.54 Natural light and space, 2005.
Source: Photographs of the author.

Natural and artificial light must be combined to improve the level of visual comfort based on the activities that are intended to be carried out in a space in a given time and the sensations that are to be obtained directly or indirectly (Figure 1.55), the use of natural light is important for energy saving and comfort, but it should not be forgotten that natural light equals heat, so it should be avoided that it enters directly in critical orientations, producing unwanted glare that generates visual discomfort to users.

FIGURE 1.55 The interior of Conran's design in Vienna, 2008.
Source: Photo by the author.

The orientation criteria based on sunlight, start from the heating of a facade depending on two aspects:

- The radiant energy you receive;
- The air temperature results in comfort.

It is about orienting the building so that it receives the most radiation in the coldest months, but that is the minimum in hot season, so it is necessary to generate shade with natural or artificial elements that are part of the environment or of the building itself.

The orientation of the buildings will depend on:

- Local topography;
- Privacy;
- Taking advantage of the views;
- Reduction of unwanted noise;
- Macroclimate and the microclimate.

In the Athens letter of 1933, second part of the requirements section, point 26, it is indicated that:

> *Studies showed that solar is essential for human health and that, in certain cases, it could be harmful. The sun is the motor of life. Medicine has shown that where the sun does not enter, tuberculosis settles; it requires placing the individual back, as far as possible, in natural conditions. The sun must penetrate every home for a few hours a day, even during the less favored season. Society will not tolerate entire families being deprived of sun and condemned to languish. Any building plan in which a single house is oriented exclusively towards the north or deprived of sun by the shadows projected on it, will be rigorously condemned. Builders must be required to provide a plan that shows that during the winter solstice the sun penetrates all homes for at least two hours a day, without this, the building license should be denied. Introducing the sun is the new and most imperative duty of the architect.*

What is stated by the Athens letter, remains as a mere recommendation, since there is no follow-up or regulation that indicates it, another aspect is that there are people who do not like the sun, even their rooms keep them in the dark, there are special cases, but even so it must be sought that all spaces have the minimum percentage of light, it is up to each user to use it or not. As Robles (2016) points out, the occupants of a space interpret light based on their experience (Figure 1.56).

FIGURE 1.56 The interior of Casa Batló in Barcelona, 2006.
Source: Photograph of the author.

The Technical Measurement Standard in Baubiologie SBM (2015) reports the following:

> *The temperature of artificial light should be as close as possible to daylight, during the day the light should be cooler, and warmer in the afternoon. The higher the color temperature, the greater the blue component should be in the light; the lower the red component, the higher. The blue and red components are the determining factors in directing the wake/sleep rhythm. Melatonin is the main responsible hormone that is governed by this element; the bluer, the lower the release of the "sleep hormone," with the redder components, the higher its release. Midday light has a very high component of blue and the evening sun has more of red (pp. 22).*

Melatonin increases below 500 luxes (lx), the ranges of this parameter vary according to the source and this is also related to the activities in the spaces (Figures 1.57 and 1.58; Table 1.13).

FIGURE 1.57 Natural lighting in a Romanian church, 2007.
Source: Photographs by the author.

FIGURE 1.58 Natural and artificial lighting in a dining room in La Pedrera, 2004.
Source: Photographs by the author.

TABLE 1.13 Lux Levels Based on Source

Light Condition	LUXES	Result
Sunny summer day	100,000	
Cloudy summer day	30,000	
Sunny winter day	20,000	
Cloudy winter day	10,000	
Gray winter day	5,000	
Clear workplace	1,000	
Lighting of a room or office	100–500	Melatonin release
Street lighting	10–50	Melatonin release
Candlelight one meter	1	Melatonin release
Full moon night	0.2–1	Melatonin release

Source: Own elaboration; Technical Measurement Standard in Baubiologie SBM (2015).

1.7.2.6 HUMIDITY CONTROL

Moisture is water vapor present in the atmosphere. Humidity control is important in a space, humidification is adding water vapor to increase the level of humidity in the air and it is used when the percentage of humidity is very low, and dehumidification is when it is necessary to eliminate the vapor of water in a climate with a high percentage of humidity; it depends on the type of local climate to select the strategy.

Accordingly, dehumidification is necessary in summer when outdoor humidity levels are high. Even when there are spaces where it is not possible to control humidity, due to different sources of production that are necessary, such as in the area of showers, washing dishes, clothes, and cooking, in addition, the same human being contributes to a certain percentage of moisture when sweating and breathing (three to five liters of water vapor per day). Sterling Ray (1980) points out, the need to dehumidify must be determined on an individual basis, since the sense of personal comfort is subjective.

An environment with a high humidity rate negatively affects the well-being and mood of the occupants, given the bad smells and the poor quality of the air that is breathed. The best solution to avoid condensation humidity is to provide adequate ventilation. If the spatial distribution does not allow natural cross-ventilation, mechanical ventilation will be required to maintain the living space with quality air. Other alternatives are to keep the kitchen and bathrooms closed to prevent the steam from spreading to other spaces, but if the kitchen is an open space, the action will be to close the other spaces and try to circulate the air through the openings avoiding blocking these air vents with other elements.

In cold weather, heating should be kept at a low level for a fairly long period, so it is important to keep the spaces ventilated with constant air renewal (Siber, 2016). It is important to have a sufficient flow of air from outside, which allows the expulsion of stale air from the space, so it is crucial to enable a natural, mechanical or hybrid ventilation system depending on the situation, which guarantees the expulsion of pollutants generated by breathing, cooking, and other sources.

The humidity in buildings is due to (Siber, 2016):

1. **Condensation:** It occurs when you have more than 70% relative humidity and you find a surface with a lower temperature that can be walls, windows, ceilings, and wood, resulting in dark stains. The

causes are an inadequate ventilation, poor insulation in the windows, a poor construction system that allows infiltrations or an inadequate heating system that does not distribute the homogenic air in the space. To avoid this, the thermal insulation of the enclosure must be increased, thermal bridges controlled, guarantee the correct renovation of spaces, use an extractor in kitchen and bathroom areas that generate humidity due to its use, in addition to calculate and install heating and cooling equipment is suitable for a good distribution of air in the space.

2. **Capillarity:** It is the humidity that rises on surfaces that have contact with the ground, for this it is important to protect them with waterproofing materials, especially if you have knowledge of the groundwater in the area.
3. **By Lateral Infiltration:** Which occurs in semi-buried constructions and which is close to a phreatic level. To avoid this, waterproofing is not only required, but is also combined with ventilation and drainage installation.

In a building, the ability to dampen temperature variations depends on the materials because of their hygroscopic capacity, the furniture and the layers that cover the surface, the fact that a construction is breathable is the ability it has to interact with the outside and the conditions that occur in different periods and that is also related to heat transfer, since it is the energy released or absorbed by its latent heat capacity, affecting the comfort, health, and energy efficiency of the building, this is called passive humidity regulation through systems based on bioclimatic strategies.

The excess humidity in interiors is generated by daily activities, mainly in the spaces classified as humid that are the kitchen, bathrooms, and the laundry room, and even by the occupant's own perspiration. The ability of materials to dampen humidity oscillations depends on their thickness, permeability to water vapor and their capacity for storing humidity. According to their microporous structure (Gómez, 2016), this is called phase change or change of state, is the fourth heat transfer mechanism, in which a substance receives or delivers thermal energy instead of changing its temperature, it changes from one physical state to another, without modifying its chemical nature (Huelsz, 2014), for what should be a parameter to consider when selecting materials for construction.

1.7.2.7 NOISE POLLUTION

Boulangeot (cited by Bueno, 1998), indicates that the transmission of noise inside buildings occurs in two ways:

- The air, which goes directly or indirectly from the emitter to the receiver, crossing obstacles being absorbed to a greater or lesser extent, according to the characteristics of the materials (Figure 1.59); and
- The pathway, where the airborne sound wave transmitted by solid bodies (an example is a blow to the wall) which, when produced, diffuses into the air.

FIGURE 1.59 A conference room, 2019.
Source: Photograph of the author.

The main objective is to isolate itself from noise, complying with two principles which is to prevent transmission, favoring isolation or prevent reflection by materials that absorb sound (soft absorbent), the denser and heavier a material is, the greater capacity insulation will have. Protecting from outside sound is very different from damping noise generated from inside.

There are different ranges of noise, from low to very high or harmful to humans, it is considered that in residential areas or hospitals, it should be maintained between 25 decibels (dB) to 40 dB, but even a normal conversation between two people can reach 60 to 65 dB, if people are speaking low or whispering, 40 to 20 dB are achieved, it also depends on the tone of people's voice. There are urban regulations that recommend less than 80 dB.

It is advisable not to exceed 30 dB at night, but a peak of 40 dB is acceptable, although 30 dB can disturb the sleep of a sensitive person. Due to each activity the levels vary (Table 1.14).

TABLE 1.14 Noise Level (Reference Values)

Levels in Decibels (dB)	Origin	Objective Feeling	Subjective Feeling
160	Toy gun near the ear	Risk of perforation of the eardrum	Pain
125 to 140	Aircraft, shotgun firing	Almost intolerant	Pain
130	Jet engines start at 50 meters	Pain threshold	Big hassle
110 to 125	Truck engine, runway, sirens, explosion	Very irritating	Big hassle
95 to 110	Dance music and motorbike motor, drill, horn, disco, shooting	Very irritating	Big hassle
80 to 95	Car engine, industry noise, traffic, bell, rail traffic	Irritating	Annoying
65 to 80	Noise from washing machine, vacuum cleaner, noisy traffic	Little irritating, stress	Tolerable
60 to 70	Noise of the day, street traffic, call, loud music	Little irritating, stress	Tolerable
50 to 65	Radio music, office, loud talk, door knock	Little irritating, borderline stress	Soft
35 to 50	Conversations in offices or public places, crowded room, lively conversation, radio, and television	Little irritating	Soft
30 to 40	Living room, quiet conversation, room volume	Little irritating to quiet	Soft
20 to 35	People chatting quietly, library, dripping faucet, rain, the clock ticking	Silent	Pleasant to annoying if constant
20 or less	Birdsong, very soft outside sounds	Silent	Pleasant
10 to 20	Quiet bedroom, wind, whisper	Silent	Pleasant
0 to 10	Breathing	Audible threshold	

Source: Own elaboration; Molar Source (2014); and Technical Measurement Standard in Baubiologie SBM (2015).

The factors that must be considered to achieve acoustic comfort are (Molar, 2014) in Table 1.15:

- Sound pressure level;
- Type of noise;
- Characteristics of the receiving subject;
- Characteristics of the task or activity;
- Characteristics of the place or space.

TABLE 1.15 Acoustic Environmental Quality Inside the Home

Interior Space	Daytime Period (7 to 22 h)	Night Period (22 to 7 h)	Observations
Bedrooms	35 to 40 dB	20 to 30 dB (although it is difficult to measure below 35 dB)	The control will depend on the orientation of the intimate area, its proximity to an external source and the material of the envelope, requires more control to achieve this level.
Living rooms	40 to 45 dB	35 dB	It is a space where certain noises can be accepted.
Service areas	50 to 55 dB	40 dB	Noise generator.
Common zones	50 decibels	There is no range	Noise generator.

Source: Own elaboration based on Molar (2014).

The environment plays an important role to achieve the adequate decibels inside the buildings (Table 1.16).

TABLE 1.16 Recommendations for an Acoustic Environmental Quality of the Environment

Source	Place	By Day (dB)	By Night (dB)
Traffic noise	Near streets and railways near residential area	59	49
	Mixed zones	64	54

Source: Own elaboration based on the Technical Measurement Standard in Baubiologie SBM (2015).

There are 10 decibels of difference between day and night, which are recommended to maintain and ensure that the interior of the houses do not have noise pollution that harms health. But this is sometimes not possible to achieve, there are subdivisions built very close to sound sources that are loud and constant.

A healthy young person hears frequencies from about 20 Hz to 20 kHz, and especially the medium ones between 1 and 5 kHz. Infrasound and ultrasound are called low and high frequency sound events below 20 Hz and above 20 kHz, which are no longer perceived by the ear, but are felt by many people, often unpleasantly that may affect people's health as indicated in the Technical Measurement Standard in Baubiologie SBM (2015).

1.7.3 EMOTIONAL ASPECT

This is sometimes omitted when designing, especially when it is produced in series, it is forgotten that people are different and that each one has an emotional and psychological need to cover, the role of a space works indirectly to generate sensations (basic and immediate experiences) and perception (which is perceived through the senses and interpreted) of what surrounds them, resulting in a positive or negative reaction in people. The psychology of color, natural, and artificial light in combination with the shape of the space or elements of the surroundings must be considered, since these can evoke different memories in each person and provoke a perception of openness or narrowness (Figures 1.60 and 1.61).

FIGURE 1.60 A zenithal light, color, and regular form, 2005.
Source: Photos by the author.

FIGURE 1.61 Natural light from the window, organic shapes in slabs and textures, 2006. *Source:* Photos by the author.

The use of color in spaces is not only an art, it is a science, you must know the tones of artificial and natural light at different times of the day and how it behaves in each space, both indoors and outdoors, as you can appreciate in the works of great architects like Barragán, the classic yellow space in one of their houses where the light at the entrance gives a feeling of warmth and mystery, for some and for others the feeling could be another. At the same time, the materials used in combination with these elements play a role in awakening the senses, influencing the connection of the inhabitants with the space they inhabit (Figures 1.62 and 1.63).

FIGURE 1.62 Natural light in the morning outside a house, 2020.
Source: Photographs of the author.

FIGURE 1.63 Artificial light in a closed space, 2009.
Source: Photographs of the author.

Natural light should not only be considered in terms of savings, but also in terms of the mood of those who inhabit it, so as not to increase stress and improve their mood when carrying out an activity, leisure, or simple contemplation.

The materials not only surround or limit a space, but also play a role in the sensations that can be generated, visual, and sound, the materials used in the surfaces of the surroundings can absorb sound, muffle it or result in a refraction, some will have attributes, even give off odors, natural materials such as wood and earth have that particularity, but currently there are new artificial materials that combined with organic waste can even smell like a drink, for example, a block made from bottles of plastic mixed with coffee husks gives off that typical smell, which for some will give a distinctive and pleasant tone to the space. There are even surfaces that keep the smell of what is done constantly in a particular space; therefore, spaces sometimes become protagonists that give an esthetic quality through the senses of the user who inhabits it (Figure 1.64).

FIGURE 1.64 The interior of a patio in a house in Saltillo, Coahuila, 2019.
Source: Photograph of the author.

Therefore, it depends on the baggage of each one, their culture, knowledge, experiences, and their tastes what makes each individual different even within the same group, which for someone may be comfortable and not comfortable for others. The word house is not the same as home since the latter implies the sentimental factor (Figure 1.65). This is also part of inhabiting, by valuing the space through experiences, the sense of appropriation of making it their own and an extension of their being, representing the personality of the person who inhabits the space leaving its essence (Figures 1.66 and 1.67).

FIGURE 1.65 The interior space of a house in Saltillo, Coahuila, 2019.
Source: Photograph of the author.

FIGURE 1.66 An intime space of a house, 2006.
Source: Photograph of the author.

FIGURE 1.67 A Moroccan Synagogue, 2005.
Source: Photograph of the author.

The way of living in a space generates symbols in architecture, responding to the context and environment of the occupant or the person who appreciates it, even when it is from another point (Figures 1.68 and 1.69).

FIGURE 1.68 A Friedensreich construction in Vienna, 2008.
Source: Photograph of Macedo.

FIGURE 1.69 Pedrera de Gaudi in Barcelona, 2007.
Source: Photograph of the author.

The experience of space has as its protagonist, the person who inhabits it (Figures 1.70 and 1.71) who builds mental symbols, giving meaning and value to the things that surround him, obtaining emotions and feelings as a result, which constitute a scenario where architecture is involved, where each space, color, texture, light, and smells will communicate a message that will be a reflection of their identity. The principle of living in semiotics is, every human being intends to build a home where the sense of appropriation, shelter, and comfort are not limiting (Robles, 2016) and for each individual the experience and expectations will result in different assessment levels.

FIGURE 1.70 A bridge designed by Zaha Hadid in Zaragoza for the expo ZH_2O, 2008.
Source: Photographs of the author.

FIGURE 1.71 The Central Plaza MQ in Vienna, 2008.
Source: Photographs of the author.

ACKNOWLEDGMENTS

I would like to thank the doctor Cristobal Noé Aguilar González for all the support provided in the preparation of this document and all the support staff for the carrying out of the work. I would like to thank all my students, for their time in helping carry out this investigation abut to COVID-19.

KEYWORDS

- **geobiological studies association**
- **nitrogen dioxide**
- **people with disabilities**
- **sulfur dioxide**
- **water closet**
- **World Health Organization**

REFERENCES

American Society of Heating, Refrigerating, and Air-Conditioning Engineers (ASHRAE), (2016). *Position Document on Limiting Indoor Mold and Dampness in Buildings*. Retrieved from: www.ashrae.org/File%20Library/docLib/About%20Us/PositionDocuments/ASHRAE---Limiting-Indoor-Mold-and-Dampness-in-Buildings.pdf (accessed on 21 December 2021).

Baubiologie Maes. (2015). Para Mediciones Técnicas Aclaraciones Y Complementos 5. Proyecto 5 (complemento). Institut für Baubiologie+Nachhaltigkeit IBN pp (14 y 17) PDF [acceso libre]. Recuperado de: http://www.sbm-standard.de/standard-2015%20condiciones%20SP.PDF (accessed on 21 December 2021).

Benévolo, L., (2002). *History of Modern Architecture*. Barcelona: Editorial Gustavo Gili.

Bueno, M., (1998). *The Great Book of the Healthy House*. Barcelona: Editorial Martínez Roca.

Cámara de diputados del h. Congreso de la unión. (2022). Ley General del Equilibrio Ecológico y la Protección al Ambiente (2021) artículo 3, fracción I. [acceso en línea]. Recuperado de: https://www.diputados.gob.mx/LeyesBiblio/pdf/LGEEPA.pdf consultado el 2 de marzo 2022.

Centers for Disease Control and Prevention (CDC), (2018). *We Remember the 1918 Influenza Pandemic.* [Online access]. Retrieved from: https://www.cdc.gov/spanish/especialescdc/Pandemia-Influenza-1918/index.html (accessed on 21 December 2021).

Chamber of Deputies LXV Legislature. (2020). Ley General de Cambio Climático artículo 3, fracción III. [acceso en línea]. Recuperado de: https://www.diputados.gob.mx/LeyesBiblio/ref/lgcc.htm consultado el 2 de marzo 2022.

Chamber of Deputies of the h. union congress. (2022). Ley General del Equilibrio Ecológico y la Protección al Ambiente (2021) artículo 3, fracción I. [acceso en línea]. Recuperado de: https://www.diputados.gob.mx/LeyesBiblio/pdf/LGEEPA.pdf consultado el 2 de marzo 2022.

Chaparro, L., (2020). Extinct Epidemic in Warsaw. Thus they decrease the typhus curve in the Warsaw ghetto. Public. [online access]. Retrieved from: https://www.publico.es/internacional/epidemia-varsovia-doblegaron-curva-tifus-gueto-varsovia.html (accessed on 21 December 2021).

CNDH, (2016). *The Human Right to a Healthy Environment for Development and Well-Being* (p. 19). Mexico: National Human Rights Commission.

Control Council Law N° 10. (1949). *Trials of War Criminals Before the Nuernberg Military Tribunal* (Vol. I, pp. 508–511). The medical case 1. US. Government Printing Office. [online]. Retrieved from: https://www.loc.gov/rr/frd/Military_Law/pdf/NT_war-criminals_Vol-I.pdf (accessed on 21 December 2021).

Curiosfera Historia (2019). *History of the Toilet: Origin, Inventor, and Evolution*. [online access]. Retrieved from: https://curiosfera-historia.com/historia-del-inodoro/amp/(accessed on 21 December 2021).

Department of Health and Human Services, (2005). Moho. *Centers for Disease Control and Prevention Safer-Healthier-People*. [acceso en línea]. Retrieved from: https://www.cdc.gov/mold/es/pdfs/dampness_facts.pdf (accessed on 21 December 2021).

Di Filippo, F., (2020). *Pandemic and Housing Crisis: The Lethal Connection Between the Reproduction of the Virus and the Type of Urbanization*. Argentine time bera. [online access]. Retrieved from: https://www.tiempoar.com.ar/nota/pandemia-y-crisis-habitacional-la-letal-conexion-entre-la-reproduccion-del-virus-y-el-tipo-de-urbanizacion?fbclid=IwAR1kQ_jmsuFbbtSF_JAhQTWHY7shTxyPPePcD1qq5uywu9PUdx1A7-jaKck (accessed on 21 December 2021).

Duncan, V., (2020). *The Spanish Flu Changed Housing*. The reason. [online access]. Retrieved from: https://www.razon.com.mx/el-cultural/la-influenza-espanola-cambio-la-vivienda/?fbclid=IwAR3vJnghNH0E-_ENkFG1GNDwtQVVmvqI5VHVCOPJDvHOFbguOmpCNarS1bw (accessed on 21 December 2021).

Espinosa, C., (2020). *What is Biohabitability*. [online access]. Retrieved from: https://www.cylex.es/reviews/viewcompanywebsite.aspx?firmaName=carmen+espinosa%2c+arquitectura+y+salud&companyId=12891237 (accessed on 21 December 2021).

Fadic, R. R., & Repetto, D. G., (2019). Measles: Historical background and current situation. *Chilean Journal of Pediatrics, 90*(3), 253–259. https://dx.doi.org/10.32641/rchped.v90i3.1231 (accessed on 21 December 2021).

Figols, M., (2016). Microbiological contamination. *Air Quality Guide* (pp. 46, 56). Coordination Knauf GmbH and FENERCOM. Spain: General Directorate of Industry, Energy, and Mines of the Community of Madrid.

FMBolivia, (2020). *Coronavirus: In Which Places is There a Greater Risk of Contagion*? [online access]. Retrieved from: https://fmbolivia.com.bo/coronavirus-en-que-lugares-hay-un-mayor-riesgo-de-contagio/ (accessed on 21 December 2021).

France 24, (2020). *Cholera in the 19th Century: From the Ports to the Bourgeoisie*. [online access]. Recovered from: https://www.france24.com/es/20200422-el-c%C3%B3lera-en-el-siglo-xix-de-los-puertos-a-la-burgues%C3%ADa-3-5 (accessed on 21 December 2021).

Fuentes, V., & García, J. R., (2005). *Wind and Architecture: Wind with Architectural Design Factor* (3rd edn.). Mexico: Editorial Trillas.

General Directorate of Architecture, Housing, and Land of the Ministry of Development, (2019). *Technical Building Code of Spain (CTE).* [online access]. Retrieved from: https://www.codigotecnico.org/images/stories/pdf/salubridad/DcmHS.pdf (accessed on 21 December 2021).

Geobiological Studies Association (GEA), (2020). *What is Biohabitability?* Retrieved from: https://www.geobiologia.org/es/que-es-la-biohabitabilidad/salud-y-vivienda/1 (accessed on 21 December 2021).

Gómez, I., (2016). Internal humidity damping capacity. *Air Quality Guide* (pp. 112, 136). Coordination Knauf GmbH and FENERCOM. Spain: General Directorate of Industry, Energy, and Mines of the Community of Madrid.

González, L. M., Casanova, M. C., & Pérez, J., (2011). Cholera: history and present. *Journal of Medical Sciences of Pinar Del Río, 15*(4), 280–294. [online]. Retrieved from: http://scielo.sld.cu/scielo.php?script=sci_arttext&pid=S1561-31942011000400025&lng=es&tlng=es (accessed on 21 December 2021).

Government of Mexico, (2020). *The New Normal: Strategy for the Opening of Social, Educational, and Economic Activities (PDF).* [online].

Guixeres, J., Higuera, J. L., & Montañana, A., (2016). Towards an emotional design in architecture: benefits in sanitary spaces. In: Aguirre, F., (ed.), *The Interior Space and the User: Theory and Interior Design* (pp. 141–158). Chihuahua, Mexico: Autonomous University of Ciudad Juárez.

Higuero, T., (2016). The importance of indoor air quality. *Air Quality Guide* (pp. 15, 16). Coordination Knauf GmbH and FENERCOM. Spain: General Directorate of Industry, Energy, and Mines of the Community of Madrid.

Huelsz, G., Molar, M. E., & Velazquez, J., (2014). *Heat Transfer in the Architectural Envelope and in the Human Being* (pp. 7–23). CA Coordinators of Technology in Architecture. Housing 2. Mexico: Autonomous University of Coahuila.

Huguet, G., (2020). Threats of humanity. *Great Pandemics in History*. National Geographic. [online access]. Retrieved from: https://historia.nationalgeographic.com.es/a/grandes-pandemias-historia_15178/1 (accessed on 21 December 2021).

IMSS. Government of Mexico, (2020). *Course All About the Prevention of COVID-19.* [online]. https://coronavirus.gob.mx/capacitacion/ (accessed on 21 December 2021).

Indoor air quality (IAQ), (2016). *Guide for the Control of Humidity in the Design, Construction, and Maintenance of Buildings*. U.S. Environmental Protection Agency. [online]. Retrieved from: https://espanol.epa.gov/sites/production-es/files/2016-07/documents/moisture_control_guidance_spanish_april_2016_508_final.pdf (accessed on 21 December 2021).

Kampf, G., Dodt, D., Pfaender, S., & Steinmann, E., (2020). Persistence of coronaviruses on inanimate surfaces and their inactivation with biocidal agents. *Journal Hospital Infection, 104*, 246–251. [online]. Retrieved from: https://www.journalofhospitalinfection.com/article/S0195-6701(20)30046-3/fulltext (accessed on 21 December 2021).

Last Minute, (2020). *How Long Does the COVID-19 Virus Last on Each Surface*? [online]. Retrieved from: https://www.ultimahora.com/cuanto-tiempo-dura-el-virus-del-covid-19-cada-superficie-n2875725.html (accessed on 21 December 2021).

Letter from Athens. Retrieved from: https://www.conservacion.inah.gob.mx/normativa/wp-content/uploads/Documento2991.pdf (accessed on 21 December 2021).

Maroto, P., (2016). The role of building materials in indoor air quality. *Air Quality Guide* (pp. 30, 36). Coordination Knauf GmbH and FENERCOM. Spain: General Directorate of Industry, Energy, and Mines of the Community of Madrid.

Measurement Technical Standard in Baubiologie SBM, (2015). *Framework Conditions* [online].

Medline Plus, (2020). *Typhus*. US National Library of Medicine, [free access] Retrieved from: https://medlineplus.gov/spanish/ency/article/001363.htm (accessed on 21 December 2021).

Mexican Chamber of the Construction Industry CMIC, (2020). *Safe Return Protocol to Construction Sites (PDF).* [online].

Microsoft News, (2020). *The INS Warns Which are the Places of Greatest Risk of Contagion of COVID-19.* [online access]. Recovered from: https://www.msn.com/es-co/noticias/colombia/el-ins-advierte-cu%C3%A1les-son-los-lugares-de-mayor-riesgo-de-contagio-de-covid-19/ar-BB14uSW7 (accessed on 21 December 2021).

Ministry of the Environment (SMA), (2020). *Metropolitan Air Quality Index (IMECA)* [online]. Retrieved from: https://www.iqair.com/mx/mexico/coahuila/saltillo (accessed on 21 December 2021).

Molar, M. E., (2014). *Comfort and the Environment: A Housing Dilemma.* Mexico: Plaza and Valdes.

Official State Gazette, (2017). *Modification of the Technical Building Code*, approved by Royal Decree 314/2006.

Olgyay, V., (1998). Architecture and climate, Bioclimatic design manual for architects and urban planners. Barcelona: Editorial Gustavo Gili, S. A.

Ordoñez, G., (2000). Environmental health: Concepts and activities. PDF [free access]. *Rev. Panam. Salud Publica/Pan. Am. J. Public. Health, 7*(3), 137–147.

Reckford, J., (2009). *How to Create Healthy Living Environments* (Part 16. No. 2, p. 4). Magazine the Forum: Cover and health.

Ríos, B., (2020). *Cholera: The Great Epidemic of the 19th Century.* Infinite geography. [online]. Retrieved from: https://www.geografiainfinita.com/2020/04/el-colera-la-gran-epidemia-del-siglo-xix/(accessed on 21 December 2021).

Robles, L. J., (2016). The sensible experience of the domestic space: the semiotic function of inhabiting. In: Aguirre, F. E., (ed.), *The Interior Space and the User: Theory and Interior Design* (pp. 59–72). Ciudad Juárez: Autonomous University of Ciudad Juárez.

Sánchez, J. J., (2019). *The Birth of the Toilet.* The siphon, an s-shaped tube, was alexander Cummings' solution to prevent the smell of discarded feces from going back up the toilet drain. National Geographic History. [online]. Retrieved from: https://historia.nationalgeographic.com.es/a/nacimiento-inodoro_14927/amp (accessed on 21 December 2021).

Sánchez, L. V., (2014). *Typhus, What the Louse Leaves Us*. The New Spain. [online]. Retrieved from: https://www.lne.es/asturama/2014/01/08/tifus-deja-piojo/1524428.html (accessed on 21 December 2021).

Serrano, P., (2017). *CO_2 Concentration in Homes and its Limitation in the CTE.* [free access]. Retrieved from: https://www.certificadosenergeticos.com/concentracion-de-co2-viviendas-limitacion-cte (accessed on 21 December 2021).

Siber, (2016). *Control the Humidity Level Inside Your Home and Protect the Health of Yours* [online]. Retrieved from: https://www.siberzone.es/blog-sistemas-ventilacion/landing/

controla-la-tasa-humedad-dentro-vivienda-protege-la-salud-los-tuyos/ (accessed on 21 December 2021).

Siber, (2016). *Indoor Quality of life, Tips for Improvement.* [online]. Retrieved from: https://www.siberzone.es/blog-sistemas-ventilacion/calidad-vida-interiores/(accessed on 21 December 2021).

Siber, (2016). *What Should be the Ideal Humidity Level at Home?* [online]. Retrieved from: https://www.siberzone.es/blog-sistemas-ventilacion/cual-debe-ser-el-nivel-de-humedad-ideal-en-una-casa/(accessed on 21 December 2021).

Siber, (2017). *The Health of Buildings: A Key Element in Well-Being and Comfort.* [online]. Retrieved from: https://ventilacion.siberzone.es/salud-de-los-edificios (accessed on 21 December 2021).

Siber, (2020). *Webinar How to Apply the New Regulations for Ventilation Systems.* [online]. Retrieved from; https://register.gotowebinar.com/recording/recordingView?webinarKey=7612775924675410445&registrantEmail=bmolar60%40hotmail.com&recurrenceKey=5923122321411570189 (accessed on 21 December 2021).

Soler & Palau. Ventilation Group, (2018). *The Effects of Air Pollution on People's Health.* PDF [free access]. Retrieved from: https://info.solerpalau.com/guias-de-ventilacion (accessed on 21 December 2021).

Sterling, R., (1980). *Land and Shelter, Semi-Buried House Design.* Center for Underground Space at the University of Minnesota. Technology and Architecture, Spain: Editorial Gustavo Gili.

The Confidential, (2020). *Anticovid Houses and Schools? The Example of How Modern Architecture Stopped Tuberculosis.* Video. [online]. Retrieved from: https://www.youtube.com/watch?v=qsaZPRbh3cs&fbclid=IwAR0hantEnfLCFPXdfJJH66cjSdiQ5UG_befWz3G2hsqhMDaLzAeBq_O7uac (accessed on 21 December 2021).

The Newspaper, (2020). *How Long Does the Coronavirus Last on Surfaces*? [online]. Retrieved from: https://www.elperiodico.com/es/sociedad/20200406/coronavirus-superficies-7892236 (accessed on 21 December 2021).

The United States Environmental Protection Agency, (1993). *Map of Radon Zones.* USEPA Publication 402-F-93-013, Washington, D.C.

The University of the Andes, (2020). *When the House Fails to be a Refuge.* Webinar.

UN HABITAT (UN HABITAT), (2010). *The Right to Adequate Housing* (pp. 1–4). Geneva: Printed at United Nations. ISSN: 1014-5567. [online]. Retrieved from: https://www.ohchr.org/Documents/Publications/FS21_rev_1_Housing_sp.pdf (accessed on 21 December 2021).

UN HABITAT (UN HABITAT), (2010). *The Right to Adequate Housing* (pp. 1–4). Geneva: Printed at United Nations. ISSN: 1014-5567. [online]. Retrieved from: https://www.ohchr.org/Documents/Publications/FS21_rev_1_Housing_sp.pdf (accessed on 21 December 2021).

UNAM, (2020). *Conditions of Habitability of Homes and the Urban Environment in the Face of Social Isolation Imposed by COVID-19.* Webinar.

Union of mutuals, (2019). *Infectious Disease Prevention Manual.* [PDF free access] General Plan of Preventive Activities of the Social Security 2019. [online]. Retrieved from: https://www.uniondemutuas.es/wp-content/uploads/2019/04/Manual-prevencion-enfermedades-infectocontagiosas.pdf (accessed on 21 December 2021).

United Cities and Local Governments UCLG, (2020). *Housing after COVID-19.* Webinar.

URBAN CENTER, (2020). *Adequate Housing, the First Line of Defense Against COVID-19: UN (WHO).* [online access]. Retrieved from: https://centrourbano.com/2020/03/31/vivienda-adecuada-defensa-covid19/?fbclid=IwAR1NJAbof5p_6lnnEztWYv_l5nwZHDPo3KkmwXKdlf93rDDyvOaDxqAC9q (accessed on 21 December 2021).

Urban Planning Laboratory, (1992). *Works on Cerdá and the Expansion of Barcelona, Reading on Cerdà and the Extension of the Barcelona Plan.* Spain: Polytechnic University of Catalonia, Ministry of Public Works and Transport MOPT, Barcelona City Council.

Uribe, N., (2016). *Open Air School (Openluchtschool), 1927–1930.* Johannes Duiker and Bernard Bijvoet. [online]. Retrieved from: https://proyectos4etsa.wordpress.com/2016/06/27/escuela-al-aire-libre-openluchtschool-1927-1930-johannes-duiker-y-bernard-bijvoet/(accessed on 21 December 2021).

Villena, J. J., (2018). *You Have an Invisible Enemy at Home, Open the Window!* Meteored. Time. [online access]. Retrieved from: https://www.tiempo.com/noticias/divulgacion/tienes-un-enemigo-invisible-en-casa-se-llama-co2.html (accessed on 21 December 2021).

World Health Organization (WHO), (1946). *How Does WHO Define Health*? [free access]. Retrieved from: https://www.who.int/es/about/who-we-are/frequently-asked-questions (accessed on 21 December 2021).

World Health Organization (WHO), (2015). *WHO Manual on Indoor Radon* (pp. 44–67). Geneva: WHO Editions.

World Health Organization (WHO), (2016). *Radon and its Effects on Health.* [online]. Retrieved from: https://www.who.int/es/news-room/fact-sheets/detail/radon-and-health (accessed on 21 December 2021).

World Health Organization (WHO), (2018). *Air Quality and Health.* [online]. Retrieved from: https://www.who.int/es/news-room/fact-sheets/detail/ambient-(outdoor)-air-quality-and-health (accessed on 21 December 2021).

CHAPTER 2

Urban Thermal Environment: Adaptation and Health

GONZALO BOJÓRQUEZ-MORALES

Faculty of Architecture and Design, Autonomous University of Baja California; Blvd Benito

Juarez, University Unit, Mexicali Baja California, CP, Mexico, E-mail: gonzalobojorquez@uabc.edu.mx

2.1 INTRODUCTION

Human beings who live in cities spend more than 90% of their time indoors. Nonetheless, for an adequate assimilation of nutrients from food, and healthy physiological and mental health, it is necessary to carry out outdoor activities. This is a clear indicator that the use of urban spaces requires adequate conditions for a longer stay in them (Elnabawi and Hamza, 2019).

Leandro-Rojas (2014), states that it is necessary to consider public space as an element of mental and physiological health. This contextualizes the criticisms of traditional models treating illnesses, with an emphasis on the consumption of drugs. The concepts of hedonic and *eudaimonia** well-being and of public space are defined. It presents the characteristics of public spaces that favor well-being, such as diversity, accessibility, and an active life, the context of some recent findings. It emphasizes active mobility as a promotor of healthy behaviors in a vibrant and diverse public space. It concludes that public spaces favor: flexibility, sensitivity to other people's suffering, curiosity, interested attention, openness to new situations and a safe context to use personal potential and energy.

Architecture for Health and Well-Being: A Sustainable Approach. Maria Eugenia Molar Orozco, PhD (Ed.)

The climate determines the thermal environment conditions in the city's urban spaces. The effect this has on the inhabitants implies specialized analysis which allows the limiting conditions of physiological health to be established and estimates the subjects' thermal adaptation in order to evaluate potential risks due to climatological variables (Figure 2.1).

Establishing the users' thermal perception of this type of space, contributes to decision-making that permits conditions between thermal comfort or acceptable. In a warm climate, as in the case study (with a historical maximum of 54°C, and maximum averages of 48°C), it is important to evaluate the subjects' adaptation level, as well as their health conditions, to avoid morbidity problems and mortality caused by heatstroke or coldstroke.

There are different researchers' interests, which have been publicized, relating to studies on thermal comfort in the urban environment (or outdoor spaces), that are analyzed in a summarized manner hereafter.

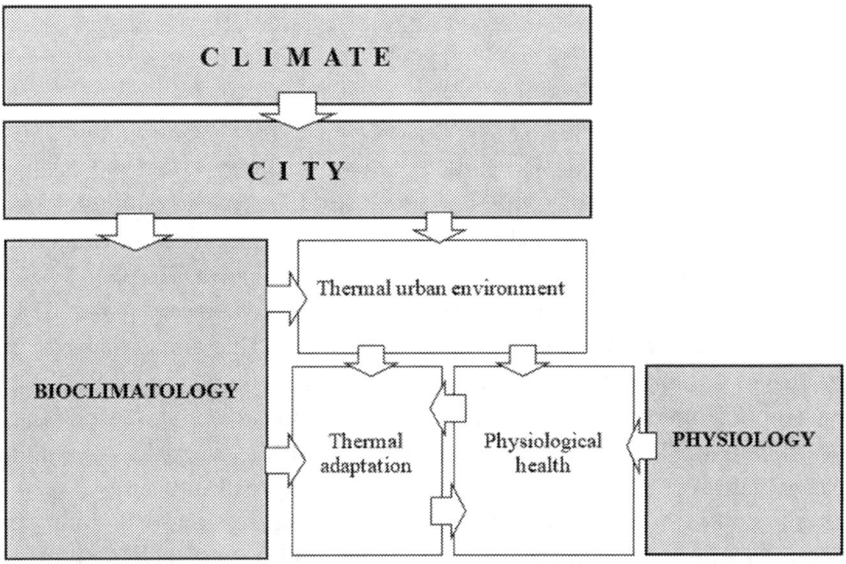

FIGURE 2.1 Interaction of components of the urban thermal environment.
Source: Elaborated by author.

Kumar and Sharma (2020) conduct a detailed review on the studies of exterior thermal comfort. They analyze the level of impact, the study methods and implementation of their results. Research completed within the last 20 years were reviewed; the predictive and adaptive approaches were considered, emphasizing the most recent advances and their contribution to the area. This research highlights that the advances in knowledge that have been made, should be implemented through the norms of urban design.

Cheng Etal (2020), establish that thermal comfort is essential for evaluating an urban spatial environment, owing to its microclimatic conditions. They carry out their research in an urban park in Mianyang, China, during the summer (with average maximums of 30°C). They registered climatological logs and applied thermal sensation surveys based on the equivalent physiological temperature (EPT). These results are presented as design proposals so as to improve the microclimate, allowing a longer use of the space to carry out outdoor activities.

Elnabawi and Hamza (2019) state that the use of outdoor spaces benefits habitability, sustainability, health, well-being, in addition to reducing energy consumption caused by the use of indoor spaces. They indicate that it is necessary to develop interdisciplinary works that allow the integration of physical, physiological, psychological, and social parameters; results which clearly assist urban planners' and designers' decision-making. A comprehensive method is suggested for the evaluation of quantitative and qualitative parameters which link the microclimatic environment with subjective thermal assessment and social behavior; the aforementioned to contribute to the development of thermal comfort standards for outdoor spaces.

Barbulescu and Lafargue (2016), establish that a city's identity is determined based on the people in its public spaces and streets; thus, making it an important study topic in a multifunctional and multidisciplinary way. Social coexistence in urban spaces requires areas with appropriate thermal comfort conditions, a component in the quality of life. With this, it will be possible to increase the number of users and the time of use. Each outdoor urban space requires specific climate adaptation strategies in accordance with the immediate microclimate.

Even though the reviewed works make important contributions to the subject, in none of the cases do they involve an analysis of both, physiological conditions and thermal adaptation, based on the perceived thermal sensation, which is the main purpose of this study.

A theoretical analysis of the urban thermal environment and health was carried out, as well as an analysis on thermal comfort in the urban environment. Later, fieldwork was conducted in a desert climate (Mexicali, Baja California, Mexico), with 2,192 surveys on users with three activity levels. The neutral temperature was analyzed based on the thermal environment, the internal body temperature (IBT), and the mean skin temperature (MST).

Subsequently, a comparative study was made between the results obtained, which demonstrate the psychophysiological and physiological adaptation levels of the subjects studied.

2.2 METHOD

The work method for the adaptation to the urban thermal environment analysis is made up of four main sections: (i) research characteristics; (ii) urban thermal environment and health; (iii) thermal comfort in the urban environment; and (iv) neutral temperature and physiological temperature. Their components are presented in Figure 2.2.

2.2.1 RESEARCH CHARACTERISTICS

Outdoor spaces' thermal environment is continuously variable, consequently, the permanence in them does not present a homogeneous pattern. Under this premise, it was considered that the study approach should be that of adaptation and not of prediction (Urías-Barrera, 2019). In this context, the work was based on a correlational analysis, applying surveys on perceived thermal sensation based on ISO 10551: 2019. The research was a transversal type divided into two periods, these considered transitional due to their temperatures: (i) March–April, transition from cold to warm; (ii) October–November, transition from warm to cold. The results obtained are presented as one due to the similarity in the results obtained regarding the perceived thermal sensation.

The variables analyzed were: (ii) dry bulb temperature (DBT), due to its psychophysiological effect on the perceived thermal sensation; (ii) internal body temperature (IBT); and (iii) mean skin temperature (MST), the latter two because of its physiological effect on thermal comfort and direct reference to thermal adaptation. The instruments being used were selected for precision, measurement ranges, and availability. A heat stress monitor was used for DBT, an infrared internal temperature thermometer (ear thermometer) was used for measuring the IBT, and the MST was measured with an infrared external temperature sensor.

The selection of the variables and instruments being analyzed was made based on the review of seven case studies on thermal comfort outdoors (Oliveira and Andrade, 2007; Hwang and Lin, 2007; Nikolopoulou, 2004; Spagnolo and deDear, 2003, Pickup and deDear, 2000; Potter and deDear,

2000; Bojórquez-Morales et al., 2012) and the ISO 7730 (2005), ISO 7726 (1998) and ISO 10551 (2019) standards. The psychophysiological and physiological effect on the perceived thermal sensation was considered.

The process implemented and the instruments used comply with the ISO 7726 (1998) requirements, therefore the data generated is Class I, according to the Brager and DeDear (1998) classification. The survey was designed based on the ISO 10551 (2019) standard and on the analysis of three thermal comfort study surveys (Nikolopoulou, 2004, Gomez-Azpeitia et al., 2007, Bojorquez-Morales et al., 2012).

The study was carried out at the Recreation Center *Juventud 2000,* located in Mexicali, Baja, California. Mexicali is a city in northwestern Mexico, located at a latitude of 32°39'54" N and longitude of 115°27'21" W, and with an altitude of four meters above sea level. The climate is extremely hot dry, with average temperatures of maximum 42°C (with extreme maximums of 49°C) and average minimum temperatures of 8°C, (with extreme minimums of –3°C) (Luna, 2008).

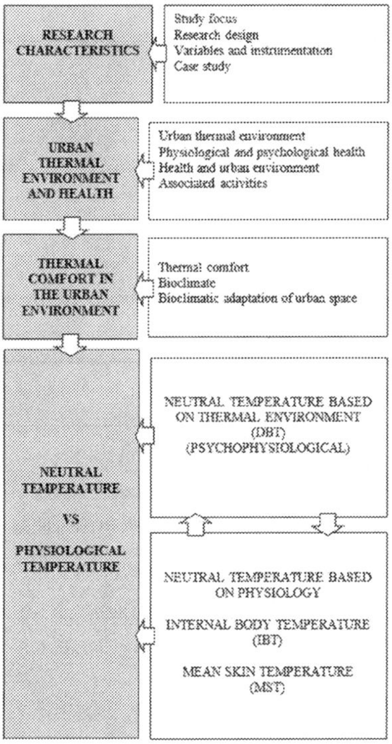

FIGURE 2.2 Urban thermal environment: Adaptation and health.
Source: Elaborated by author.

2.2.2 URBAN THERMAL ENVIRONMENT

The urban thermal environment analysis was the theoretical basis that allowed the compression of the effect of human thermal adaptation for the study conditions. For this, it was divided into two sections: (i) urban thermal environment and health; and (ii) thermal comfort in urban thermal environment.

In the urban thermal environment and health section, a study was conducted. This study goes from conceptualization, and components of the term, to the analysis of the effect this type of space has on the users' mental and physiological health, establishing exactly how important it is. A description of the activities that can be carried out, and how this influence the quality of life in the city, is also highlighted.

Thermal comfort in the urban thermal environment is analyzed; its main components and the interrelation of the space itself with psychophysiological processes and their effect on human thermal comfort, and on perceived thermal sensation and adaptation to environmental conditions.

2.2.3 ADAPTATION AND HEALTH

The quantitative estimation of the subjects' adaptation and health was based on a correlational study between the perceived thermal sensation compared with DBT, IBT, and MST. The periods studied were: (1) November to October (from the warm period to the cold period), and (2) March to April (from the cold period to the warm period). The study dates were: April 1 to 13, 2008, October 27 to November 9, 2008, and April 13 to 26, 2009, from 07:00 to 21:30.

The sample design was based on the number of people who go to the recreational park in the selected periods. A sample was designed, with a reliability of 95% and estimators' precision of 5%. The value obtained was 380 observations on average per period. Attributable to the acceptance of the study among the respondents, a total of 2,192 observations were reached, which include: 1,284 of passive activity (75 W/m^2, mean value), 485 of activity moderate (183 W/m^2, mean value) and 423 of intense activity (600 W/m^2, mean value).

The measuring instruments were checked before doing the fieldwork and applying the surveys. The subjects, to whom the surveys were applied, were selected randomly, but because of the poor response rate, it was proceeded to do so in a deterministic way. The study subjects were men and women between the ages of 12 and 65, individuals with irregular biological conditions such as temporary or chronic diseases, pregnancy, lactation period or menstrual period were not included. The surveys were conducted during the park's operating hours (06:00 to 22:00, Monday to Sunday), in such a way that all periods of use were covered in relation to the level of attendance at the site.

The neutral temperature estimate (as a reference for human thermal adaptation) for DBT, IBT, and MST was carried out with the method of means by thermal sensation interval (MTSI) (Gomez-Azpeitia et al., 2007), which is based in Nicol's (1993) proposal for "asymmetric" climates. With the MTSI, unlike the conventional method, before obtaining the regression line, strata are determined to calculate the mean value and standard deviation (SD) of each one of them, therefore, the regression is not done with all the pairs of data in the sample, but only with the mean values and the ranges are established by adding and subtracting one or two times the SD.

This procedure is used to determine the mean temperature value of all responses for each perceived thermal sensation level. The value of the subjects' mean temperature, not only of those who said they felt thermally comfortable, but also of those who expressed other thermal sensations, is calculated. The field data was processed separately according to each of the seven thermal comfort response categories as per ISO 10551 (2019).

In order to establish the human thermal adaptation level, based on the psychophysiological and physiological evaluation, a comparative analysis was made between activity levels according to type of neutral temperature studied: (i) thermal environment temperature (DBT); (ii) internal body temperature (IBT); and (iii) mean skin temperature (MST). The aspects studied were: the neutral value (comfort temperature), thermal comfort ranges and determining coefficient. It is worth mentioning that the comparisons were made, in all cases, based on the value of the thermal sensation of comfort for the passive activity level. Lastly, the types of neutral temperature were compared based on their mean neutral values and amplitude ranges.

The information analysis regarding the perceived thermal sensation was a mixed, qualitative type from a phenomenological perspective, initially based on the regression lines, through the development of the assumptions that were observed in the graphic results. A quantitative analysis was also carried out based on the characteristics of the regression equation of the mean regression line for DBT, IBT, and MST.

2.3 URBAN ENVIRONMENT AND HEALTH

When talking about urban environment, reference is made to a set of terminologies such as: 1) Environment which, in general terms, is defined as a set of physical and biological factors that have a direct or indirect impact on a living organism and influences its development and 2) Urbanism, which refers to the study and planning of cities, linked to a guided system, a set of interrelated rules and principles.

Within urban planning, there are several terms that allow a better perception of what the urban environment refers to. There is the term urban equipment, which is defined as a set of services, either for industries, or in this case, urbanizations. Urban equipment offers services that generate physical and mental well-being of those who make up a community. Ghysais, (2018), refers to the function of urban equipment as a collective use, such as squares, parks, theaters, universities, among others (Figure 2.3).

FIGURE 2.3 Squares as collective equipment. (a) "Juventud 2000" shopping mall. Mexicali, B.C. Mexico; (b) school campus square UABC, Vice-rectory. Mexicali.
Source: Photograph of the author.

Another definition, within the category of urban environment, are public spaces. Quijije and Castro (2020) cite the United Nations for Human Settlements UN-Habitat (2018) program, which defines public space as places of public property, free or non-profit, that are accessible and enjoyable for the community, and these include both streets, as well as open spaces and public facilities. In addition, they emphasize its importance for social and individual well-being.

Therefore, for the urban environment, all the terms mentioned above converge to define it as the set of circumstances and physical elements (Figure 2.4), generated by a city's urban growth and development, dedicated to the community, that affect the physiological and psychological spheres of the individual.

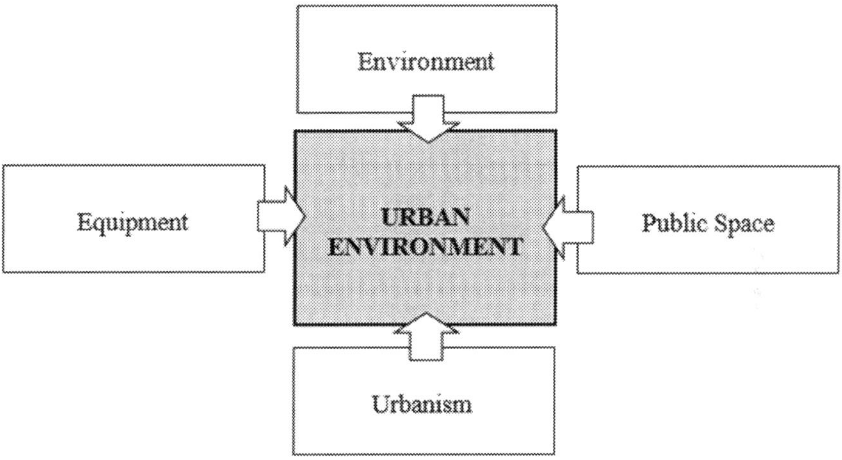

FIGURE 2.4 Terminologies in relation to the urban environment.
Source: Elaborated by author.

2.3.1 URBAN ECOSYSTEM SERVICES

Within the urban environment, there are two concepts called ecological infrastructure and urban ecological services. The first one talks about the elements that make up an ecological system in a city, and they are understood through natural, semi-natural, and artificial networks. The second refers to the services provided by an ecological infrastructure to benefit, directly or indirectly, the community's inhabitants (Karis, Magali, and Ferraro, 2019).

An urban ecological service can be classified into three categories: 1) Supply services or product supplier that come directly from ecosystems such as water and food. 2) Regulation or maintenance services in ecosystems such as decomposition, water filtration or the climate and 3) Cultural services that, unlike the previous two, are intangible because they are obtained from a direct and indirect interaction with nature.

As far as urban green areas, such as open public spaces, parks, and gardens, the category to which it belongs to is cultural services, since it underlines outdoor activities like sports or recreational activities, and they are classified within this area as an urban ecological service because they provide long-term benefits to human health and well-being. Green areas are a system that provides goods and services to the community that inhabits them (Jennings Etal, 2015) (Figure 2.5).

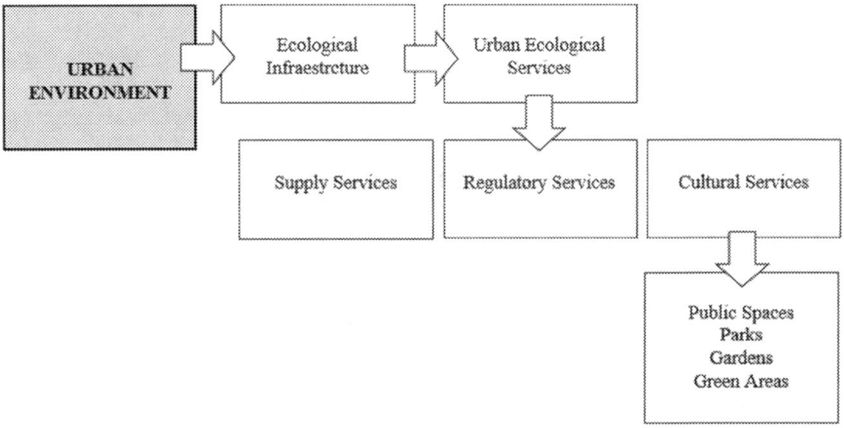

FIGURE 2.5 Relation between urban environment and urban ecological services. *Source:* Elaborated by author.

There is evidence of the benefits or advantages that the urban ecological supply and regulation services, mentioned above, can provide, such as improving the environment from air and water pollutants, mitigating urban heat effects, and improving the access to foods such as fruits and vegetables.

The urban environment's ecological services contributions are considered as tangible or material benefits. In the same way, they offer non-material benefits, because of the impacts they generate are not tangible. These benefits are promoted by cultural ecological services and their elements such as the esthetics of the landscape; the values that are generated both spiritual and cultural, promote sports, physical, and recreational activities contribute to the physical and psychological well-being of the human being (Karis, Magali, and Ferraro, 2019).

The benefits in the urban ecological system of the cultural service are related to the exposure time that the user has in these areas. As Aerts Etal (2018) describes, a short exposure to urban parks or semi-natural environments reduces symptoms of stress, depression, and fatigue, and increase positive emotions and feelings of well-being. And for prolonged or "chronic" exposures, where the user resides in easily accessible areas, or within a distance between 150 meters to 5 km, the benefits increase in measure, associating them to the reduction of mortality from conditions such as cancer, respiratory, and cardiovascular diseases.

Among other advantages, crime incidence is lessened, by providing quality public spaces. Because, otherwise, having, a deteriorated public space in the urban environment, such as one with insufficient lighting, promotes the feeling of insecurity, negatively affecting the community, and that consequently counteracts the motivation to practice outdoor activities.

2.3.2 PHYSIOLOGICAL HEALTH

The urban environment is closely related to the health of the individual who lives in the community. The urban environment contains physical and biological factors that affect the user who partakes in their environment. As Galarraga et al. (2018) cite, "The urbanized and built environment is part of the determinants of health."

There is an intermediary between the relation that health and the urban environment have, which is called urban environmental quality, where Quijije and Castro (2020) cite Luengo (1998), who describes it as the product of the optimal conditions' interaction of the habitable space, in terms of comfort, to bring forth the formation of a healthy habitat capable of satisfying the basic needs for both individual and social sustainability.

As far as public health is concerned, the urban environment provides a wide range of benefits that address essential issues about diseases that must be treated. For example, Chee, Jordan, and Horsley (2015) explain the existence of global epidemics such as obesity, cardiovascular, and respiratory problems, which largely affect the population given the increasing numbers of cases, as a result of the lifestyles, directed to a sedentary way of life, that have been adopted.

To address this matter, government agencies and health professionals have chosen to reduce this problem through the implementation of public spaces, such as green areas, that encourage the community to engage in outdoor activities. Similarly, they state that the conditions of the urban environment are directly related to physical activity, being that they are linked to the relation between availability and accessibility to open public space. In other words, the mere existence of a good quality public space within the urban environment has a considerable influence on society in carrying out physical activities that help reduce the risks of suffering from diseases.

For the elderly sector of the population, having green areas and public spaces at their disposal, improves their physical condition. As they perform their outdoor activities, such as walking in their free time, it helps them face the conditions or limitations caused by health problems like chronic diseases.

Chang (2020) explains how the quality of the urban environment, such as parks, green areas, and walkways, plays an important role in elderly persons' health. He classifies the preferential activities for this sector, such as relaxation, walking, interaction with plants, interaction in social groups, and physical exercise. In his study, he divided the participants with the age of 75 or more, who have modified their styles of activities due to mobility losses, adjusting them to perform only relaxation activities. The other group, those under the age of 65, who have an active lifestyle, participated in outdoor activities with more frequency. The influence of a good quality of urban environment and a correct planning of these spaces in order to improve the quality of life in the sector stands out.

This also stands out in the children population sector, where there is evidence of the health benefits caused by outdoor spaces for their development. In this stage, humans have a greater and faster learning process, so the environment in which he/she grows takes an important role in its development. As Aziz and Said (2015) describe, children games are mechanisms that are part of their development and help them to recognize and learn from their environment; therefore, playing outdoors stimulates their senses and cognitive abilities. There are different types of outdoor environments within the urban environment in which the child grows, and each one affects his/her perception, development, and evolution in a different manner (Table 2.1).

There are multiple benefits in children's health so as to avoid and prevent the development of diseases. Urban outdoor environments and green public spaces are considered indispensable for prenatal development, since exposures to these have positive effects on the newborn's weight. Exposure, in the newborn's first stage of development, reduces the risk of developing diseases such as schizophrenia and atopic sensitization. As far as long-term effects, being in contact with the environment's beneficial microbiome has effects on the development of the immune system (Aerts Etal, 2018).

2.3.3 PSYCHOLOGICAL HEALTH

Just as in physical health, mental health is also positively affected by the urban environment. Mental health is a set of social, emotional, and psychological factors that nourish the individual's feeling of well-being.

The World Health Organization (WHO, 2018) highlights, among other data, three points that define mental health. First, it elevates the condition of mental health to a condition of greater relevance than to that of one with just the absence of mental disorders. Second, it explains that mental health is an integrated part of health in general, because if there is an absence of mental well-being, there is not a good state of health. And third, it describes that mental health is determined by socioeconomic, biological, and environmental factors. This last point relates the topic addressed, affirming the influence that the urban environment has on the individual's health.

TABLE 2.1 Urban Environments and Child Use

Types of Urban Environments	Characteristics
Neighborhood	It can be the house yard or its surroundings, such as a park. It offers the opportunity for a greater frequency of activity and independent development.
School play area	It promotes physical activity, offering different game options that help improve the child's imagination, awakening interests and abilities, facilitating social and cognitive interaction.
Playground	It contains thought-provoking elements for the child, giving him/her the opportunity to manipulate, explore, and experiment, while developing cognitive awareness, social, and motor skills.
Streets	Streets within the neighborhood, or those that are not too busy, is a preference on the child's behalf for the wide range of activities offered.
Public spaces	Well-designed public spaces can promote children's interest and participation in the city.
Green areas	A diverse natural landscape has the ability of satisfying children's needs, being a stimulating space that promotes a variety of games and activities.

Source: Elaborated by author based on Aziz and Said (2015).

According to studies performed, there is evidence of the power that open public spaces in urban environments has on the mental and psychological health of the individual (Wood et al., 2017). They conducted a study on the impact open public spaces and green areas have on mental health. Their results showed that as the number of parks and gardens increased in the neighborhood, an increase in good mental health in the community was also perceived, because people had direct contact with the space, walking through it, or an indirect contact by surrounding the space, "restoring" their mental well-being.

Another important factor that relates the user's psychological state to outdoor spaces is the feeling of satisfaction; If the quality of the cultural environment (parks, gardens, and public outdoor areas), meets the standards so that the activities are carried out in a fluent manner, the level of user satisfaction rises and, therefore, that of the entire community. This in turn, increases the levels of attachment to the neighborhood and the connection between the user and nature is strengthened by promoting the development of an eco-centric and pro-environmental behavior that helps benefit both human beings and wildlife (Southon Etal, 2018) (Figure 2.6).

FIGURE 2.6 Neighborhood community "Colonia Nueva" doing outdoor activities, Mexicali, B.C., Mexico.
Source: Author's photograph.

In addition to the terms green spaces or green areas, which refer to an open space with vegetation, there are also the terms blue spaces or blue areas, which refer to all waters surfaces visible within the space, such as lakes, rivers, fountains, among others. These are mentioned to a lesser extent, but like green spaces, they fulfill the role of improving the quality of mental health, since they reduce the risk of depression and psychological distress, with just simply seeing them.

2.3.4 ACTIVITIES

Performing outdoor activities, essentially, depends on having appropriate urban spaces. It also depends on factors such as climate and temperature as much as work and leisure hours. In cities such as Mexicali, Baja California, whose climate is extremely hot and dry, the climate-temperature factor is closely related to the frequency and the hours in which the inhabitants decide to carry out these physical activities outdoors.

When performing any type of physical or sports activity, the body temperature increases as a result of the heat production derived from metabolism. Consequently, if the environmental temperature is higher than the body temperature, and if it is not possible to dissipate the heat produced by the metabolism, you run the risk of provoking a heat stroke. Another factor that interferes with users being able to perform physical activities is in the type of recreational spaces in which they take place. As mentioned before, the quality of the public space, as well as the accessibility, size, design, and proximity of green and recreational areas, influence the user's decision-making in regard to performing these outdoor activities (Tables 2.2–2.4).

Since this is related to human thermal adaptation, a study of activities in urban recreational spaces (directly related to physiological and psychological health) was conducted, as a way of determining the user types, schedules, and ages of the people involved. The spaces analyzed are representative to the city and have a capacity that represents 20% of the population of the municipality of Mexicali, Baja California, Mexico.

The study shows that the predominant activities are sports and exercise; the last one being social interaction. It is worth mentioning that the Vicente Guerrero Park was not designed for the use of sports and exercise, even though these activities are performed there (Table 2.2).

TABLE 2.2 Users' Frequency by Hours and Type of Activities Performed in Four Urban Recreational Spaces in Mexicali, Mexico

Recreational Space	Months/Days/Hours with Greatest Frequency	Type of Space Required	Users' Activities and Distribution
Integral Development Center "Centenario"	• Months: April–August • Days: Tuesday and Thursday • Hours: Morning, afternoon, and evening	• Green areas • Sport areas • Recreational areas	• Sports (50%) • Exercise (20%) • Social Interaction (10%) • Recreation (10%)
Sports Unit at UABC Campus	• Months: June–August • Days: Tuesday, Wednesday, and Thursday • Hours: Afternoon and evening	• Green areas • Sport areas • Recreational areas	• Sports (65%) • Exercise (20%) • Social Interaction (10%) • Recreation (5%)
Recreation Center "Juventud 2000"	• Months: March–November • Days: Tuesday, Wednesday, and Thursday • Hours: Morning, afternoon, and evening	• Green areas • Sport areas • Recreational areas	• Sports (75%) • Exercise (15%)
Vicente Guerrero Park	• Months: January–May and October–December • Days: Saturday and Sunday • Hours: Morning and afternoon	• Green areas • Sport areas • Recreational areas	• Sports (20%) • Exercise (15%) • Social Interaction (30%) • Recreation (35%)

Source: Elaborated by author.

In reference to the time and age distribution (Table 2.3), it is observed that it varies from one space to another, depending on the housing development types and the predominant families in them. For example, in areas close to residential areas already consolidated, with more than 20 years of foundation, it is observed that young people and adults predominate, with the presence of few children. While, in areas close to new housing developments, children, and adolescents are the predominant groups. There are also differences depending on the hours, whereas greater attendance is observed in the morning, there is less at noon and at night.

TABLE 2.3 Distribution by Time and Age in Recreational Spaces, in Mexicali, Mexico

Distribution by Time and Age	**Morning 6–11 h**		**Afternoon 12–18 h**		**Night 18–22 h**	
	Age	**%**	**Age**	**%**	**Age**	**%**
	Integral Development Center "Centenario"					
	Children	10	Children	35	Children	15
	Adolescents	30	Adolescents	25	Adolescents	30
	Young adults	30	Young adults	20	Young adults	30
	Adults	25	Adults	15	Adults	25
	Elderly adults	5	Elderly adults	5	Elderly adults	5
	Sports Unit at "Universidad Autonoma De Baja California" Campus					
	Children	10	Children	25	Children	10
	Adolescents	10	Adolescents	7.5	Adolescents	25
	Young adults	35	Young adults	30	Young adults	30
	Adults	35	Adults	30	Adults	25
	Elderly adults	10	Elderly adults	7.5	Elderly adults	10
	Recreation Center "Juventud 2000"					
	Children	2	Children	15	Children	2
	Adolescents	40	Adolescents	25	Adolescents	35
	Young adults	–	Young adults	–	Young adults	30
	Adults	40	Adults	15	Adults	15
	Elderly adults	5	Elderly adults	3	Elderly adults	1
	Vicente Guerrero Park					
	Children	10	Children	35	Children	15
	Adolescents	20	Adolescents	25	Adolescents	30
	Young adults	30	Young adults	20	Young adults	30
	Adults	35	Adults	15	Adults	25
	Elderly adults	5	Elderly adults	5	Elderly adults	5

Source: Elaborated by author.

The activity and age distribution shows greater activity in adolescents, followed by young adults, and then by children and adults, with elderly adults being the least active. This social behavior is due to the fact that the user's frequency in a public space to perform physical and recreational activities, is in accordance with the user's age and the type of activity. The time of day and means of travel also influence, just as much as the family's customs and way of life (Table 2.4).

TABLE 2.4 Distribution by Activity Type and Age in Recreational Spaces in Mexicali, Mexico

Distribution by Activity Type and Age	Sports		Exercise		Social Interaction		Recreation	
	Age	**%**	**Age**	**%**	**Age**	**%**	**Age**	**%**
	Integral Development Center "Centenario"							
	Children	10	Children	5	Children	30	Children	30
	Adolescents	20	Adolescents	15	Adolescents	20	Adolescents	15
	Young adults	30	Young adults	35	Young adults	15	Young adults	15
	Adults	25	Adults	30	Adults	15	Adults	20
	Elderly adults	5	Elderly adults	5	Elderly adults	5	Elderly adults	20
	Sports Unit at "Universidad Autonoma De Baja California" Campus							
	Children	10	Children	10	Children	10	Children	20
	Adolescents	30	Adolescents	20	Adolescents	10	Adolescents	15
	Young adults	30	Young adults	30	Young adults	20	Young adults	15
	Adults	20	Adults	30	Adults	20	Adults	20
	Elderly adults	10	Elderly adults	10	Elderly adults	40	Elderly adults	30
	Recreation Center "Juventud 2000"							
	Children	2	Children	15	Children	2	Children	2
	Adolescents	40	Adolescents	25	Adolescents	35	Adolescents	40
	Young adults	–	Young adults	–	Young adults	30	Young adults	–
	Adults	40	Adults	15	Adults	15	Adults	40
	Elderly adults	5	Elderly adults	3	Elderly adults	1	Elderly adults	5
	Vicente Guerrero Park							
	Children	10	Children	5	Children	30	Children	20
	Adolescents	20	Adolescents	25	Adolescents	20	Adolescents	20
	Young adults	30	Young adults	35	Young adults	15	Young adults	15
	Adults	25	Adults	30	Adults	15	Adults	30
	Elderly adults	5	Elderly adults	5	Elderly adults	20	Elderly adults	15

Source: Elaborated by author.

2.4 THERMAL COMFORT IN URBAN SPACE

Thermal comfort in urban space, especially in a desert climate, determine its density and use frequency. It is essential to understand the reason why a space created for recreation or coexistence is underused as a result of the thermal environment conditions. For this, it is necessary to analyze the space's bioclimate, as well as its adaptation, based on its conditioning factors, and the direct impact on the perceived thermal sensation (Figure 2.7).

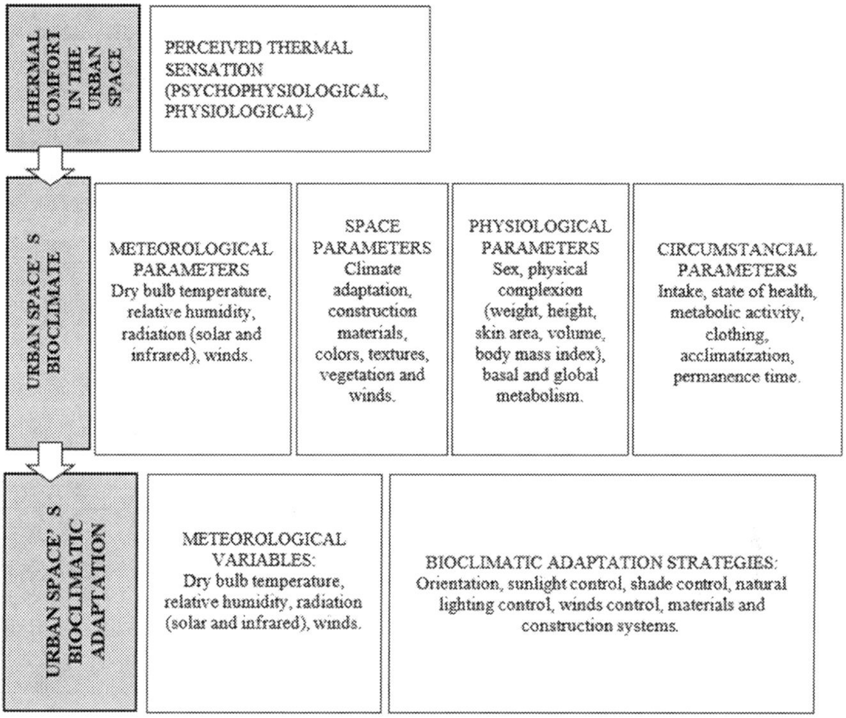

FIGURE 2.7 Conditioning factors for thermal comfort in outdoor spaces.
Source: Elaborated by author.

The bioclimate is defined based on the effect that the meteorological conditions of a thermal environment have on humans. The parameters that influence thermal comfort for outdoors, although they are similar to those of indoor comfort, are found within a wider range of variability, temporality, space, and activities. Because of this complexity, there have been few attempts to understand outdoor thermal comfort (Höppe, 2002).

The urban space's adaptation will depend, primarily, on the predominant meteorological variables for which it should be designed, aiming on having between comfortable and bearable conditions. In order to achieve this, bioclimatic adaptation strategies should be implemented based on:

orientation, sunlight control, shade control, natural lighting control, wind control, materials, and construction systems.

2.4.1 THERMOREGULATION

The internal temperature which is considered normal, without affectations, oscillates around 37.6°C, in an interval of 36°C to 38°C; however, during vigorous physical activities it can reach 40°C, which, in specific circumstances, is necessary to achieve adequate performance. An essential health condition is to maintain the internal temperature within the limits of ± 4 or 5°C (Mondelo et al., 2001).

The MST can vary, depending on: clothing, environmental conditions, and metabolic activity. This temperature establishes the skin's ability to transfer heat to the environment. The internal or central temperature (from the deep tissues of the body), is the weighted average of the different temperatures of the parts and organs of the body. These temperatures vary according to the activity, body part and time. They oscillate with the circadian rhythm and remain approximately between ± 0.6°C (except when there is a fever); even if the individual is exposed to low temperatures of 12°C, or high temperatures of 60°C.

The hypothalamus monitors and controls the body's internal changes, and regulates blood pressure and body temperature. It transforms emotions to responses or physical changes and is decisive in the thermal exchange, between the parts of the human body and the outside (Mondelo et al., 2001). From a physiological point of view, this organ establishes the perceived thermal sensation; however, there are psychological effects that affect the body's response to the thermal environment (Figure 2.8).

The human body anticipates changes in internal temperature based on the perception of the environment's temperature and the level of activity it performs. The sensation of thermal discomfort is generated due to the fact that the internal organs need to function at a temperature of 37°C. When the thermal environment conditions modify it, adaptation mechanisms are activated, consciously or unconsciously, intending to maintain it.

There may be environmental conditions in which thermal equilibrium is unavailable for a long time or when the metabolism stops working properly to the point that the temperature of the internal organs drops to 30°C (hypothermia) or rises to 41°C (hyperthermia). In both situations death occurs.

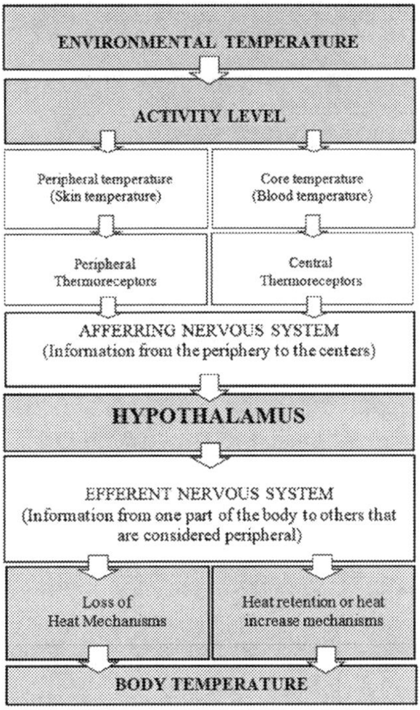

FIGURE 2.8 Thermoregulatory system.
Source: Own elaboration.

Physical activity is important in increasing heat production (approximately 20%). During an intense level of activity, the metabolic rate can increase 20 to 25 times above the basal level, which theoretically increases the internal temperature by 1°C every five minutes (McArdle et al., 1990). This generates an energy exchange with the thermal environment that determines the effect on the thermal sensation that is perceived.

The human body's mechanical efficiency is low, between 75% and 100% of the energy it consumes and produces to perform activities is converted into heat within its body, depending on the type of activity, to which must be added the heat produced by the basal metabolism necessary to stay alive (Mondelo et al., 2001).

The human body anticipates changes in internal temperature based on the environmental temperature perception. This reading is done through two sets of sensory organs within the skin, which are the Krause's end-bulbs body (sensitive to heat loss) and the Ruffini's end organs (sensitive to heat gain) (Urias-Barrera, 2019).

Heat sensory impulses regarding the cold ones, are delayed to try to avoid false records, because increasing or decreasing the IBT takes a long time. If the heat and cold sensors signal it simultaneously, the brain inhibits one or both of the body's defense reactions. If the body temperature rises too much, the vasodilation process begins, in which blood flow through the skin increases and as a consequence begins to sweat to reduce the temperature, since the energy required by sweat to evaporate is taken from the skin.

On the other hand, if the body lowers its temperature too much, the blood vessels react with vasoconstriction that generates an erection of the hair follicles, which in turn reduces the flow of blood through the skin and increases the thermal resistance of the skin. Another reaction is to increase the production of internal heat that generates muscle convulsions, whose purpose is to increase the metabolic heat (tremors or shivering). The continuous generation of metabolic heat does not always guarantee the minimum internal temperature necessary for living and for carrying out activities when people are exposed to certain cold conditions, so low temperatures can be dangerous. However, generally speaking, high-temperature environments are more dangerous than cold ones, since it is usually easier to protect oneself from the cold than from the heat (Urias-Barrera, 2019).

Inbar Etal., (2004) show how the thermoregulation system's efficiency is directly related to age and sex. A younger person has better heat dissipation or retention response. In addition, there is the effect of the individual's skin area proportion and type of clothing. Although there have been studies on temperature and relative humidity effect on human thermoregulation, according to Kim and Jeong (2002), there is an influence the lighting level on the thermophysiological reaction, which is also affected by the level of clothing. and skin exposed to light.

2.4.2 THERMAL SENSATION AND ADAPTATION

The thermal sensation represents the final phase of the thermal environment analysis process. The process begins with a perceptual reading of the environment, where psychological and physiological aspects are involved. Later this first information is analyzed, from which the thermoregulation process, previously described, begins (Figure 2.9).

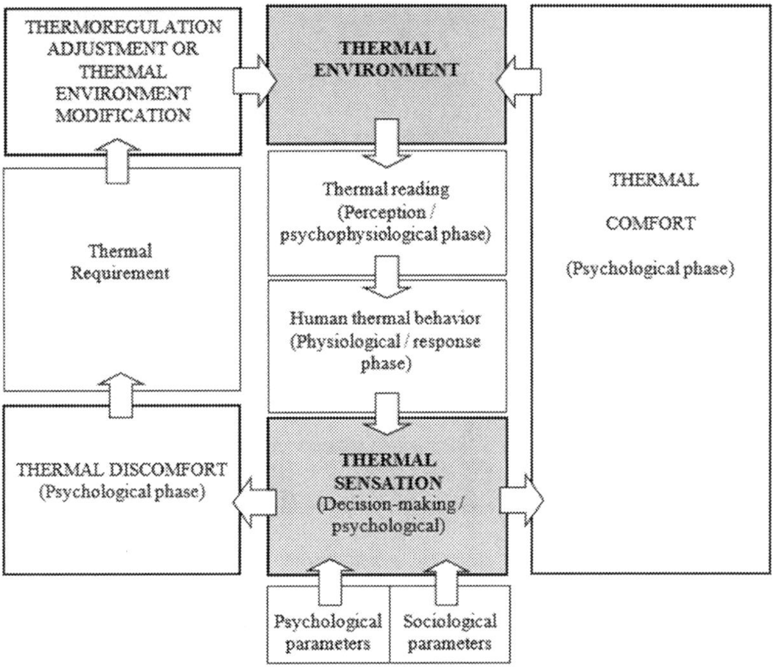

FIGURE 2.9 Thermal sensation perception process.
Source: Elaborated by author.

TABLE 2.5 Parameters of the Perceived Thermal Sensation

Type of Parameter	Components
Meteorological	Air temperature, relative humidity, wind speed, solar radiation, and infrared radiation.
Built space or natural space	Materials, colors, textures, artificial air conditioning, climate adjustment, vegetation.
Circumstantial	Activity (metabolic relation), clothing, state of health, acclimatization, time spent in the thermal environment, intake.
Physiological	Sex, age, physical complexion (weight and height: skin area, body mass index), basal metabolism, global metabolism.
Psychological	Personal habits and preferences (behavior), expectation, and experience, control of the source of discomfort, environment stimulation, reactive adaptation (behavioral), interactive adaptation (behavioral).
Sociological	Cultural environment, social environment, lifestyle, customs, diet, social status, nationality.

Source: Elaborated by author.

Based on the thermal sensation, it is observed that any action taken to restore a comfortable thermal sensation that has been altered, must be understood as part of a process that has two fundamental sections: (i) the one that collects and transmits the factors' information that determines the thermal conditions (internal or external) to the brain; and (ii) the one that processes this information, determines a reaction and transmits it to the different organs of the body in order to carry it out (Gomez-Azpeitia et al., 2007).

Some thermal comfort models consider only the physiological part of the perceived thermal sensation, while others, based on the adaptive approach, include, inherently, although not explicitly, psychophysiological aspects. There are simple indices that try to estimate the perceived thermal sensation from two variables. There are others that are more complex and include metabolism, clothing insulation and up to five more thermal environment variables in their analysis. It is important to mention that the selection of a thermal comfort model depends on the study's purpose. However, if the information needed for its application is not available or is complex to use, the use of simple models is encouraged.

Human thermal adaptation represents the greatest influence on responses generated in the hypothalamus regarding perceived thermal sensation. The thermal comfort sensation is attained through the user's perception depending on his/her degree of adaptation and the variables that influence in reference to the space. According to Nikolopoulou and Steemers (2003), thermal adaptation is considered as: "the gradual decrease of the organism's response to repeated exposures to stimuli from a specific environment." The adaptation is divided into physiological and psychological. The components of this classification are presented in Table 2.6.

Human thermoregulation responds to stimuli external to the body, but when these vary in type and intensity constantly (solar radiation, wind, temperature, relative humidity), the gradual decrease in response is less than when the stimulus is almost constant, despite it being of various types (temperature, and relative humidity, wind, and radiation emitted by the surroundings). Acclimatization is achieved if there is a gradual decrease in the effect of the stimulus and it reaches a minimum.

The space's naturalness is defined by the presence of materials and furniture that modify the natural condition of the space. According to Nikolopoulou and Steemers (2003) it is increasingly evident that people can tolerate more changes in a natural physical environment.

There are aspects, such as experience and expectation, that present a significant variability from one outer space to another, since their meteorological conditions change constantly, so it is not possible to compare the experience from one space to another. This does not happen for interior spaces since, according to their uses, they have similar general characteristics.

TABLE 2.6 Variables of Thermal Adaptation

Variable	Description
Physiological Adaptation	
Acclimatization	The intensity and type of stimuli can vary according to the thermal environment. If there is a gradual decrease in the effect of the stimulus and it reaches a minimum, acclimatization is achieved.
Psychological Adaptation	
Space's naturalness	Environment without artificiality. It is increasingly demonstrated that people can tolerate major changes in the physical environment in a natural way.
Expectation and experience	Expectation: what the user assumes the environment should be, even more than what it is. Experience: memory effects of sensations and experiences in other spaces. They relate these to "new spaces" which directly affects people's expectations.
Exposure time	An exposure to discomfort is not recognized as negative if the individual anticipates the situation to which he/she will be exposed to and controls the exposure times. Exposure times can be established based on the activity being performed, or the condition of discomfort itself.
Perception control	People who have a high degree of control and influence over the source of discomfort, tolerate wide variations of it and the user's negative emotional responses are reduced, because they can restore that comfort.
Environmental stimulation	When a thermal environment becomes intolerable it is rejected by the user. The environment stimulates the user and he/she establishes under their specific conditions whether or not they are in thermal comfort.
Reactive behavior	Changes that are part of personal behavior such as: type of clothing, posture, and physical position or metabolic alterations through the consumption of hot or cold beverages.
Interactive behavior	Behavioral actions in which people make decisions and modify the immediate thermal environment to achieve conditions of thermal comfort.
Sociocultural environment	The lifestyle and local customs such as types of construction, diet, clothing, and social coexisting activities, in addition to empirical knowledge of the climate, influence the experience and thermal expectation, therefore, the perceived sensation.
Habits and preferences	Personal habits and preferences influence the thermal preferences of individuals. These would vary depending on the type of activity: passive, moderate, and intense.

Source: Elaborated by author.

The sociocultural environment refers to the set of extrinsic conditions of the lifestyles and customs, and the empirical knowledge of the climate, which determines the types of buildings, clothing, and forms of social coexistence, which in turn influence the thermal sensation expectation, as a result of several generations' activities. It establishes behavioral patterns, based on the lifestyle and local customs that impact clothing, food, use, and design of spaces, which in turn affects the subjects' conditions and thermal environment. Based on the above, the influence on the psychological adaptation of the cultural environment where people grow is noted. In addition, there is a noticeable influence on the type of activities, time, and period of use of the spaces, which affects the acclimatization process.

Personal habits and preferences influence the thermal preferences of individuals. These vary, depending on the type of activity (passive, moderate, and intense) and the sociocultural environment, according to the uses of the space. On the other hand, they affect the level of physiological and psychological adaptation, and in turn, in the experience and thermal expectation in reference to the conditions of the spaces used.

When the organism is exposed to consecutive and similar physical activities in a determined thermal environment, adjustments are generated in the psychological and physiological mechanisms of thermoregulation between the first four to seven days. Mondelo et al. (2001), affirm that acclimatization in hot climates can vary from seven to 14 days, also pointing out that for each rest day, half a day of acclimatization is lost.

According to Rhoades and Tanner (1997), the acclimatization indicators appear in the first days of heat and exercise exposure combined and most of the tolerance improvement occurs in approximately 10 days. Acclimatization to heat is transitory; it disappears within two to three weeks after returning to a more temperate environment. In extreme climates, absolute acclimatization can take months or even years.

2.4.3 BIOCLIMATE AND ADJUSTMENT

An adequate space design implies that the users are comfortable, so that they can perform their activities not only in an adequate way, but also so that they can be performed in an integrated manner. However, most architects are unaware that, in terms of thermal comfort, there is a significant difference between the physiological and psychological reactions of individuals when it comes to outdoor and indoor spaces. The importance of the thermal comfort study of interior spaces is observed with the development of the standards ISO 7730: 2005, ANSI/ASHRAE 55: 2017, ISO 7243: 1989, ISO 7933: 2004, ISO 11079: 2007, and ISO 10551: 2019 among other.

The need for serious studies about outdoor thermal comfort conditions has been revealed in globally significant events like the Olympic Games in Atlanta (1996), Sydney (2000), and Athens (2004), Expo Sevilla World Fair (1992), along with research projects such as rediscovering the urban realm and open spaces (RUROS), as well as some ISO standards that besides being used for interior spaces have applications in exterior spaces. The contributions of these works are applicable in tourist, educational, recreational projects, or exhibition areas.

In regards to the thermal adaptation variables, which will be specifically discussed later, there are determining factors that illustrate the difference between the thermal comfort of outdoor spaces and that of indoor spaces.

In what can be considered the built thermal environment, the spatial conditions have a striking influence on the user and their thermal comfort perception. The conditions vary according to the activity and the type of space.

The variables of the constructed space that influence the thermal environment are different for outdoor and indoor spaces and are classified as: meteorological conditions, activities, hours of use, construction materials, furniture, and adaptation strategies (Table 2.7).

TABLE 2.7 Variables of the Constructed Space and Their Influence on Thermal Comfort Perception

Variables	Outdoor	Indoor
Meteorological conditions	Variation continues for 24 hours	Variation controlled by air conditioning equipment and/or partially by the architectural surrounding.
Activities	Variation by season and schedule	Defined with little variation
Hours of use	Variation by season and activity	Pre-established
Construction materials	Weather-resistant materials, sometimes materials of natural origin override	Typical materials of the region, and sometimes some from different climates to where it is built, are used.
Furniture	Made with weather resistant materials.	Pre-established materials and models commonly used to depend on the activity.
Adaptation strategies	Passive type strategies prevail and active ones rarely do.	Active type strategies prevail, even with the use of passive ones.

Source: Elaborated by author.

Exterior climatic conditions present constant variations during the 8,760 hours of the year due to the changes in temperature, relative humidity, solar radiation, speed, and direction of winds. Whereas, indoors, even when the envelope is affected by the external meteorological conditions, there is a controlled variation, of the previously mentioned variables, by air conditioning equipment or climatic adaptations of the building.

The activities that take place in outdoor spaces may vary based on the season and with the hours of use, meanwhile activities in an indoor space, have little variation, and are regularly defined by the function of the space itself, such as an office, a library, etc.

The hours of use of an exterior space, rather than an interior space, are not only different in time, but the exterior space's hours change from one season to another based on the activity to be performed. However, indoor activities will be, generally, the same every day.

As to the construction materials, it is important to make a suitable selection since they will be exposed to weather conditions and may or may not accumulate energy that will affect the thermal comfort conditions of the space.

Although the materials used to build exterior and interior spaces have some characteristics in common, exteriors, generally, are weather resistant and require little maintenance. While for interior materials the selection is based on providing an internal environmental control that can promote thermal comfort through its surroundings. For the interior finishes there are criteria such as texture, costs, or personal tastes.

The furniture is chosen according to the use of the space, but the materials used depend on its specific function and location (exterior or interior). However, the effect of the thermal load accumulated in them is rarely considered in reference to the activity use. Hence, for exteriors, a wooden bench is more suitable than a metal one in a warm climate. For interiors, the selection of furniture will affect the thermal load removed by artificial cooling and the thermal environment of the space's user.

The adaptation strategies for an outdoor or indoor space vary greatly, nonetheless, the objective in both cases is the same: to have thermal comfort conditions for the user. In outdoor spaces, adaptation strategies are generally passive such as orientation, tree planting, vegetation typology, shadows, materials, furniture, passive fans, fountains, reflecting waters, etc. In special cases, active-type evaporative or fan coolers are also used, in addition to cooling columns.

In interior spaces, the adaptations can be through the surroundings' materials, roofs, orientation, vertical shading, horizontal shading, colors, textures, thermal mass, natural ventilation, and in most cases, active cooling systems such as fans, coolers evaporative, air conditioners, among others.

The bioclimatic adaptation of outdoor spaces, considers meteorological adaptation strategies, with aspects such as: orientation, sunlight, shadows, natural lighting, natural ventilation, materials, and construction systems, for the decisions made in reference to the design proposal.

The main premise of bioclimatic design is knowing the site's meteorological variables conditions of the site where the building will be constructed, in order to control them based on the user's thermal comfort requirements.

The bioclimatic adaptation of outdoor spaces gives the inhabitants of regions with extreme climates the possibility to be in direct contact with the sun, the air and especially with other human beings. There are some types of outdoor architectural spaces such as stadiums, cinemas, and open-air theaters, outdoor exhibition areas, parks, recreational centers, school playgrounds, outdoor areas of hotels, fairs, among others, where the importance of the bioclimatic adaptation is observed so that it is possible to properly carry out the activities for which they were created.

According to Vidal (2004), in outdoor public space projects, the aspects of the users' physical-environmental well-being are not properly studied. esthetic landscape design concerns are those that predominate, and although they are not less important, the problem is that for a place to be habitable, that is, comfortable for the user's requirements in relation to a specific activity, being beautiful is not enough. A fundamental requirement is that it satisfies the environmental comfort needs, that is, thermal, acoustic, lighting, visual, and safety aspects.

In regard to these needs, it can be observed that the majority of outdoor public spaces, generally, present poor habitability conditions in terms of environmental comfort to potential users including: noise pollution, thermal discomfort, unregulated luminosity, and visual pollution.

The demand for outdoor public spaces indicates the need to externalize community values, through constructed forms, and culture values, through the uses of those constructed forms. The people who demand these spaces are potential users, that is, producers of collective values and a consumer of public space (Vidal, 2004). Due to the absence of this concern, the outdoor public space is used only to circulate, to stay for a short time, and in most cases, it is not enough to solve the thermal comfort problem. The abovementioned leads to the idea of intensive use of outdoor public spaces, depending on seasonal, meteorological, and time conditions.

The intensity of use refers to the number of users per unit of time and space. The greater the number of users, in less space with a restricted duration of time, the greater the intensity. If the number of users decreases, proportionally to the decrease in spatial unit and duration in time, the high intensity will be maintained. If the number of users decreases, while the spatial unit and the occupation time increases, it is said that there is a decrease in the intensity of use. Finally, if the increase in users is less than the increase in the spatial unit, and in the occupation time, it is also considered a decrease in intensity.

Moving from conventional architecture to bioclimatic architecture implies a shift in paradigm, since it suggests a new way of designing (with emphasis on the environmental well-being of the user) and a new way of living, which means moving from a passive user to an architecturally active environment (of sophisticated artificial air conditioning technique) to an active user in a passive environment (where the architecture design captures all the passive energy available in the elements of the climate).

To achieve this complete paradigm shift, a thorough consciousness is required, not only for need to use solar energy in order to design outdoor public spaces that allow greater intensity and duration use, but also a consensus around the idea that bioclimatization of outdoor public spaces is a possible and effective alternative.

For this it is necessary: (i) Establish regulations for designing forms, materials, techniques, and use of vegetation in those spaces, in order to guarantee environmental comfortable objectives; (ii) Establish proven methodologies and verification techniques of designs with dynamic modeling; (iii) Establish measurement and evaluation systems for public spaces; (iv) Establish environmental quality standards for outdoor public spaces; and (v) Establish scaled conditions for public spaces in relation to the level of environmental well-being goal that is wanted to be achieved.

In Nikolopoulou's (2004) work, a complete interdisciplinary analysis of the environmental conditions of outdoor spaces is presented. In addition, some passive adaptation proposals are made for areas of social coexistence, recreation, or transition. Also, in this document, the possible passive adaptation strategies in outdoor spaces based on comfort type and control variables (Table 2.8).

A common mistake is that exterior spaces' designs are carried out with methods made for interior spaces and, even more, the evaluation processes of the thermal environment conditions. The effect of variability of meteorological conditions, nor the effects of infrared radiation, hours of use, activity levels, clothing, among others, are not considered. Therefore, there is a need to have thermal comfort models that allow estimating the perceived thermal sensation based on the various characteristics of users and spaces to be carried out.

2.5 NEUTRAL TEMPERATURE VS. PHYSIOLOGICAL TEMPERATURE

The neutral temperature (Tn), was developed by Bedford in 1936. With this model the theoretical basis of the adaptation approach is born. It is an index of perceived temperature, based on surveys applied to light industrial workers in England. It considers the metabolic effect, with moderate to intense activity levels and light to heavy clothing (Auliciems and Szokolay, 1997). Even though it was designed for indoor spaces, its practicality allows it to also be used as a method for outdoor spaces.

TABLE 2.8 Passive Adaptation Techniques for Outdoor Spaces

Comfort Type	Passive Suitability Technique	Considerations
Thermal comfort	• Solar radiation and lighting control • Wind control • Humidity control • Floor construction systems • Outdoor furniture materials	• Shade caused by trees, supportive structure with vegetation as shade, horizontal or vertical elements with providing shade tor areas of circulation or rest areas. • Use of trees, shrubs, plants on the structure, vertical elements such as fences, even the building itself. • Fountains, reflecting or mirroring waters, vegetation or desiccant walls, outdoor wind towers, water micronizers. • Radiant or wet floors, vegetal or plastic finish. • Use of materials suitable for climate type, preferably wood or plasticized.
Acoustic comfort	Noise control	Use of tree barriers, support structure with vegetation, horizontal or vertical elements such as acoustic shields, earth slopes or changes in terrain level. Use of sources as noise eliminators.

Source: Elaborated by author.

For this research, the neutral temperature estimate was made for: (i) the perceived thermal sensation relative to the external thermal environment's

DBT; (ii) the perceived thermal sensation relative to IBT; and (iii) perceived thermal sensation per MST. In all three cases, it was performed for the total of observations made and an analysis for three metabolic levels. Subsequently, a comparison of the results obtained was made (Figure 2.10).

2.5.1 NEUTRAL TEMPERATURE BASED ON THE THERMAL ENVIRONMENT

The thermal sensation due to DBT, in transition periods, for the total amount of observations, showed few cases outside the limits of thermal comfort, towards the end of the greater extensive range. In reference to the lower extensive range, there were only some cases outside the regression line of -2DS.

This result was due to the fact that, in this case, the three levels of activity considered, each have different thermal characteristics and differences in the subjects' psychological adaptation (Figure 2.11(a)), as mentioned by Humphreys and Nicol (2002); and Nikolopoulou and Steemers (2003).

It is worth mentioning that during the "very hot" thermal sensation (7), a greater dispersion of the data was observed, in comparison to other thermal sensations. This represented a lower level of adaptation by the subjects to the heat sensations.

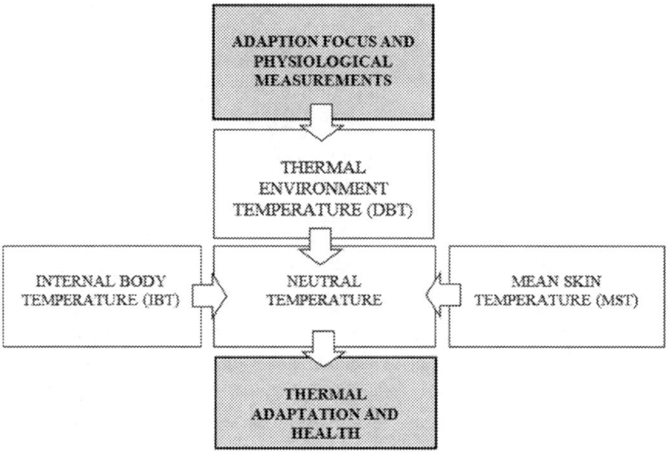

FIGURE 2.10 Thermal adaptation and health based on thermal environment, internal body, and mean skin neutral temperature.
Source: Elaborated by author.

The amplitude of ranges, both extensive and reduced, revealed significant variability in the conditions of the subjects' thermal and psychological adaptation, a situation which coincides with Humpreys and Nicol's (2002) statements. This was due to the fact that, in this case, all three levels of activity are included.

Significant differences were observed between the adaptation levels for each sensation scale of the subjects being studied. There was a greater variability as the temperature increased. This indicated a lower degree of adaptation to high temperatures in the period studied, which concurs with Humpreys and Nicol's (2002) adaptation theory. The graphic analysis revealed that the DBT thermal sensation in transition periods for passive activity has a similar behavior in all of the observations, in terms of data dispersion, regression lines and variation in the thermal sensation scale (Figures 2.11(a) and 2.11(b)).

The moderate activity, regarding the thermal sensation by DBT, did not present thermal sensation values of "very hot" (7) or "very cold" (1); It also had a data dispersion in the "hot" (6) and "cold" (2) sensations. The responses obtained were concentrated between thermal sensations of: "fairly cold," "neither hot nor cold" and "fairly hot." This indicated that most of the subjects were in comfortable and tolerable conditions, according to what was established by Fanger (1972) (Figure 2.11(c)).

It is worth mentioning that this level of activity is where the greatest dispersion of data was generated because of the different types of activities carried out and to the variability of the subjects' adaptation levels due to the discontinuity of the activities being practiced. This confirmed Rhoades and Tanner's (1997) statements.

The regression lines were (visually) parallel with each other and to the mean regression line, which indicated similar conditions in the subjects' perception regarding the thermal environment for both cold and warm conditions in the period studied and for this level of activity. This behavior also established conditions consistent with those of a symmetrical climate in relation to thermal sensations, since the values of thermal comfort towards hot or cold climates were perceived in similar conditions. This also concurred with what was published by Nicol (1993), in regard to the types of climates of thermal sensation.

A similarity was observed between the central values ("Fairly hot," "Neither hot nor cold" and "Fairly cold") in the thermal sensation scales, This, coupled with the concentration of data in these sensation scales, indicated that for this period and this level of activity the conditions of the thermal environment have a tendency towards the subjects' thermal comfort.

The thermal sensation of DBT in transition periods of intense activity, was the only activity level, within the period, that presented the sensation of "a lot of heat;" however, with dispersed values. This was due to the metabolic level and human thermoregulation, as established by Fanger (1972); and Mondelo et al. (2001) (Figure 2.11(d)).

The regression lines were convergent with the mean regression line, as the thermal sensation of cold increased, which represented a greater adaptation to cold conditions, due to the level of metabolic activity generated in intense activities, in accord with ISO 7243: 1989, as well as Mondelo et al. (2001).

There was a significant variation between the values for "Fairly cold" and "Neither hot nor cold." This was due to the internal temperature generated by the subjects while the activity was carried out (in total movement) and the change in the thermal environment at the time of the survey (static), where an inverse thermodynamic exchange occurred, in addition to the convective cooling effect, as established by Fanger (1972).

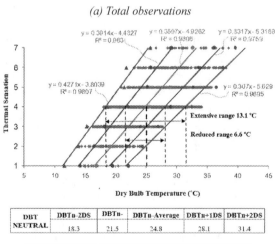

(a) Total observations

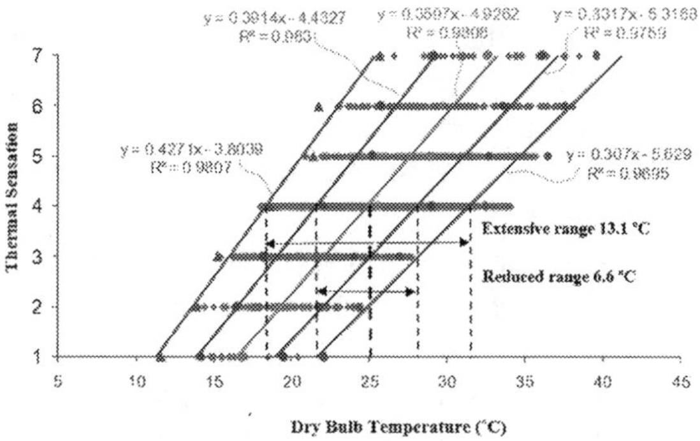

DBT NEUTRAL	DBTn-2DS	DBTn-	DBTn-Average	DBTn+1DS	DBTn+2DS
	18.3	21.5	24.8	28.1	31.4

DS	Thermal Sensation	Scale	-2DS	-1DS	Average	+1DS	+2DS
3.5	Hot	7	25.7	29.1	32.6	36.1	39.6
4.0	Warm	6	21.7	25.7	29.6	33.6	37.6
3.8	Slightly Warm	5	21.4	25.2	28.9	32.7	36.5
3.5	Neutral	4	18.6	22.0	25.5	29.0	32.5
2.9	Slighvlty Cool	3	15.2	18.2	21.1	24.0	27.0
2.6	Cool	2	13.8	16.5	19.1	21.7	24.4
2.6	Cold	1	11.5	14.2	16.8	19.4	22.1

FIGURE 2.11 *(Continued)*

(b) Passive activity

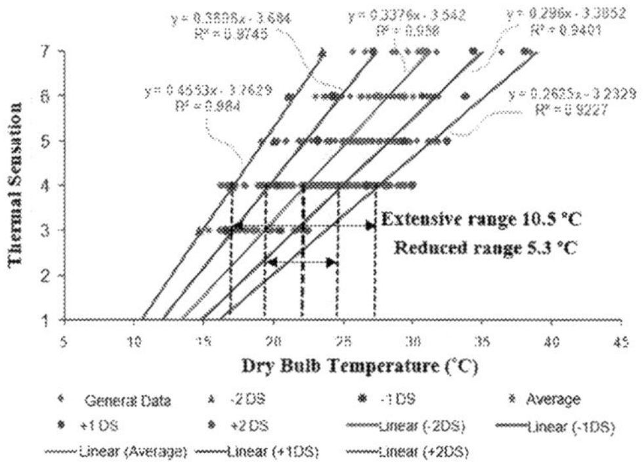

DBT NEUTRAL	DBTn-2DS	DBTn-1DS	DBTn-Average	DBTn+1DS	DBTn+2DS
	17.1	19.7	22.3	25.0	27.6

DS	Thermal Sensation	Scale	-2DS	-1DS	Average	+1DS	+2DS
3.6	Hot	7	23.5	27.1	30.8	34.4	38.0
3.2	Warm	6	21.1	24.3	27.5	30.7	33.9
3.1	Slightly warm	5	20.0	23.1	26.2	29.4	32.5
3.2	Neutral	4	17.0	20.2	23.5	26.7	30.0
1.9	Slightly cool	3	14.7	16.6	18.6	20.5	22.4

(c) Moderate activity

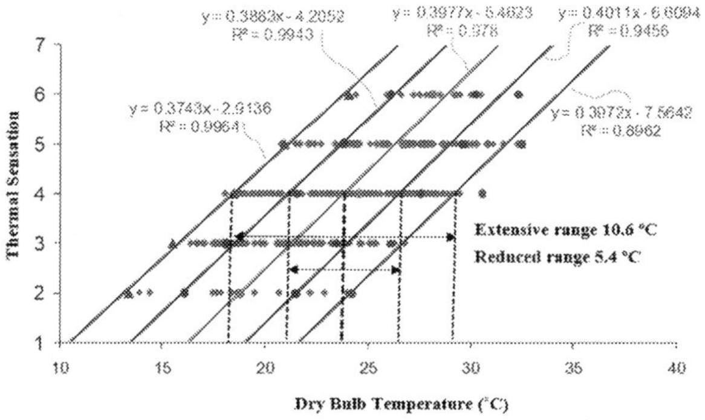

DBT NEUTRAL	DBTn-2DS	DBTn-1DS	DBTn-Average	DBTn+1DS	DBTn+2DS
	18.5	21.1	23.8	26.5	29.1

DS	Thermal Sensation	Scale	-2DS	-1DS	Average	+1DS	+2DS
2.1	Warm	6	24.0	26.1	28.1	30.0	32.3
2.9	Slightly warm	5	20.9	23.8	26.7	29.6	32.5
3.0	Neutral	4	18.5	21.5	24.5	27.5	30.6
2.6	Slightly cool	3	15.5	18.2	20.8	23.4	26.1
2.7	Cool	2	13.4	16.1	18.8	21.5	24.2

FIGURE 2.11 *(Continued)*

(d) Intense activity

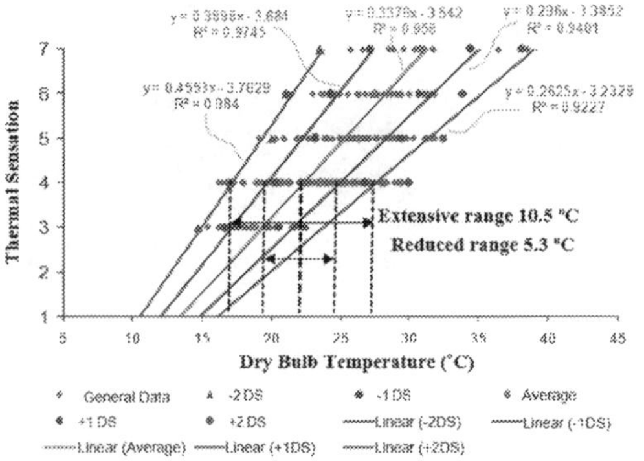

DBT NEUTRAL	DBTn-2DS	DBTn-1DS	DBTn-Average	DBTn+1DS	DBTn+2DS
	17.1	19.7	22.3	25.0	27.6

DS	Thermal Sensation	Scale	-2DS	-1DS	Average	+1DS	+2DS
3.6	Hot	7	23.5	27.1	30.8	34.4	38.0
3.2	Warm	6	21.1	24.3	27.5	30.7	33.9
3.1	Slightly warm	5	20.0	23.1	26.2	29.4	32.5
3.2	Neutral	4	17.0	20.2	23.5	26.7	30.0
1.9	Slightly cool	3	14.7	16.6	18.6	20.5	22.4

FIGURE 2.11 Neutral dry bulb temperature and thermal comfort ranges in transition periods for: total observations, activities: passive, moderate, and intense.
Source: Elaborated by author.

2.5.2 NEUTRAL TEMPERATURE BASED ON PHYSIOLOGY

Due to the importance of gathering information regarding the subjects' external and internal temperature changes, physiological aspects were considered. The thermal sensation by physiological conditions was evaluated by a correlational method based on the adaptation approach. The following were analyzed: IBT and MST. Although it was possible to estimate other variables with these variables, it was not considered adequate since they were based on equations developed under other conditions of thermal environment or subjects with different degrees of adaptation.

The thermal sensation by IBT allows us to identify the recording conditions for the decision-making of the hypothalamus in reference to the perceived thermal sensation, according to the thermoregulation aspects mentioned by Mondelo et al. (2001).

The body's internal temperature was measured using an infrared ear thermometer, whose shape is designed to enter only what is necessary in the ear canal, completely avoiding contact with the eardrum. The infrared heat of the eardrum and adjacent tissue was measured. To ensure greater accuracy, the probe tip was at a temperature similar to that of the human body.

The MST establishes the thermal interaction conditions between the human body and the thermal environment, such as: heat transfer by air (convection), radiative transmissions with surrounding surfaces (radiation) and the latent heat exchanges produced by the evaporation of sweat.

The MST was recorded with an infrared thermometer, which works by projecting two rays whose ideal distance to the measurement area is the one where both coincide on the same point of light projection.

Three measurement points were used: 1) the front and the back to the palms of the hands (in both hands); this was defined based on Krause's end-bulbs body distribution (sensitive to heat loss) and Ruffini's end organs (sensitive to heat gain), in addition to the fact that these areas generally do not have clothing (except for some cases in very low temperatures) and are usually are in direct contact with the thermal environment's conditions.

2.5.2.1 THERMAL SENSATION BY INTERNAL BODY TEMPERATURE (IBT)

In all the observations, the thermal sensation by IBT had a variation of 0.5 to 1°C of mean values (Mean), in reference to the value of thermal comfort. Regarding the minimum (± 1DS) and maximum (± 2DS) values of each level of thermal sensation, a maximum variation of 5.4°C was obtained. It was noted that there were few cases outside the estimated thermal comfort limits. The regression lines were convergent with each other as the thermal sensation of heat increased. The variation between the amplitude of the extended and the reduced range was 1.2°C. The difference in the determining coefficient (R2) for the five regressions revealed a maximum variation of 0.03 (Figure 2.12(a)).

In the passive activity's case, the thermal sensation by IBT, presented a variation of 0.4 to 0.8C for the mean values of the perceived thermal sensation level, in reference to the thermal comfort value. Regarding the estimated values for ± 1DS and ± 2DS per thermal sensation interval, a maximum variation of 4.2°C was obtained. It was observed that the regression lines of ± 1DS and ± 2DS were convergent, with each other and in reference to the TnMean line, as the thermal sensation of heat increased. The variation between the amplitude of the extended range and the reduced one was 0.8°C.

The difference of R2 for the five regressions presented a maximum variation of 0.15 (Figure 2.12(b)).

In the moderate activity's case, the thermal sensation by IBT had a variation of 0.6 to 1.3°C of mean values (TMean) of the perceived thermal sensation level, in reference to the thermal comfort value. Regarding the estimated values for ± 1DS and ± 2DS per thermal sensation interval, a maximum variation of 4.1°C was obtained. It was observed that the regression lines of ± 1DS and -2DS were generally visually parallel, however, in +2 DS's case, a convergence effect was presented when the thermal sensation diminished to very cold. The variation between the amplitude of the extended and the reduced range was 1.3°C. The difference of R2 for the five regressions presented a maximum variation of 0.06 (Figure 2.12(c)).

Meanwhile, in the intense activity's case, the thermal sensation by IBT, had a variation of 0.4 to 1°C of mean values (Mean) of the perceived thermal sensation level, in reference to the thermal comfort value. Regarding the estimated values for ± 1DS and ± 2DS per thermal sensation interval, a maximum variation of 3.2°C was obtained. It was observed that all the regression lines of ± 1DS and ± 2DS were convergent in reference to the regression line of the Mean; this happened as the cold thermal sensation increases. The variation between the amplitude of the extended range and the reduced one was 1°C. The difference of R2 for the five regressions presented a maximum variation of 0.02 (Figure 2.12(d)).

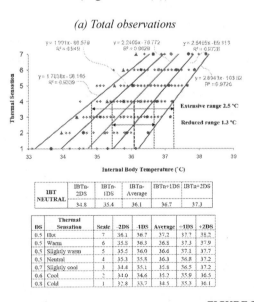

(a) Total observations

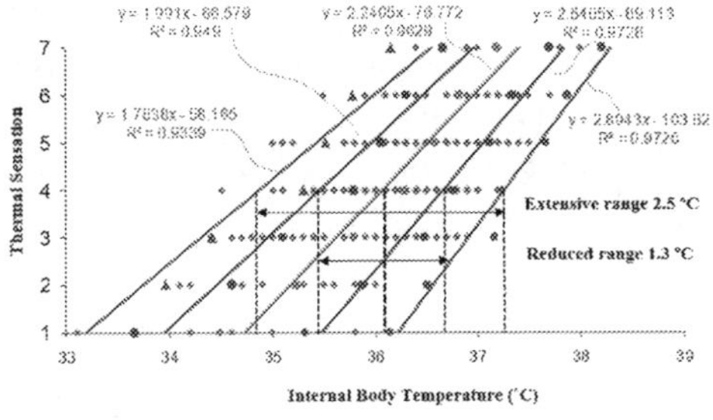

IBT NEUTRAL	IBTn-2DS	IBTn-1DS	IBTn-Average	IBTn+1DS	IBTn+2DS
	34.8	35.4	36.1	36.7	37.3

DS	Thermal Sensation	Scale	-2DS	-1DS	Average	+1DS	+2DS
0.5	Hot	7	36.1	36.7	37.2	37.7	38.2
0.5	Warm	6	35.5	36.3	36.8	37.3	37.9
0.5	Slightly warm	5	35.5	36.0	36.6	37.1	37.7
0.5	Neutral	4	35.3	35.8	36.3	36.8	37.2
0.7	Slightly cool	3	34.4	35.1	35.8	36.5	37.2
0.6	Cool	2	34.0	34.6	35.2	35.9	36.5
0.8	Cold	1	32.8	33.7	34.5	35.3	36.1

FIGURE 2.12 *(Continued)*

(b) Passive activity

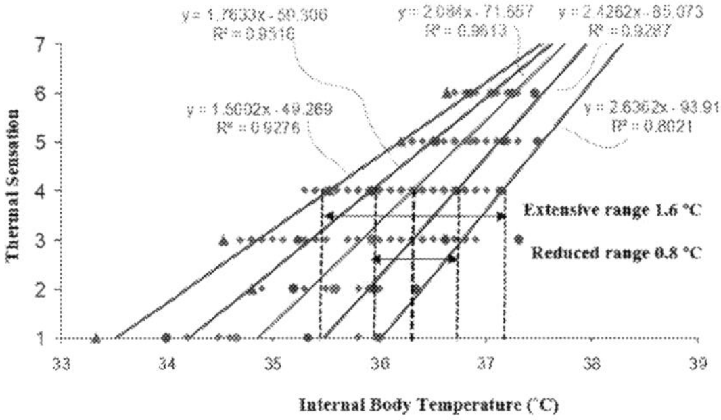

IBT NEUTRAL	IBTn-2DS	IBTn-1DS	IBTn-Average	IBTn+1DS	IBTn+2DS
	35.5	35.9	36.3	36.7	37.1

DS	Thermal Sensation	Scale	-2DS	-1DS	Average	+1DS	+2DS
0.21	Warm	6	36.6	36.8	37.0	37.3	37.5
0.32	Slightly warm	5	36.2	36.5	36.8	37.2	37.5
0.40	Neutral	4	35.5	35.9	36.3	36.7	37.1
0.69	Slightly cool	3	34.5	35.2	35.9	36.6	37.3
0.38	Cool	2	34.8	35.2	35.6	36.0	36.3
0.66	Cold	1	33.3	34.0	34.7	35.3	36.0

(c) Moderate activity

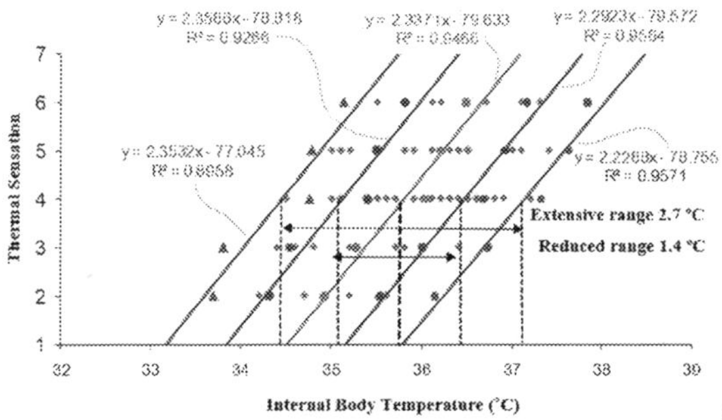

IBT NEUTRAL	IBTn-2DS	IBTn-1DS	IBTn-Average	IBTn+1DS	IBTn+2DS
	34.4	35.1	35.8	36.5	37.1

DS	Thermal Sensation	Scale	-2DS	-1DS	Average	+1DS	+2DS
0.68	Warm	6	35.13	35.81	36.48	37.16	37.84
0.71	Slightly warm	5	34.78	35.49	36.21	36.92	37.63
0.64	Neutral	4	34.76	35.40	36.04	36.68	37.32
0.73	Slightly cool	3	33.82	34.54	35.27	36.00	36.73
0.61	Cool	2	33.71	34.32	34.93	35.53	36.14

FIGURE 2.12 *(Continued)*

(d) Intense activity

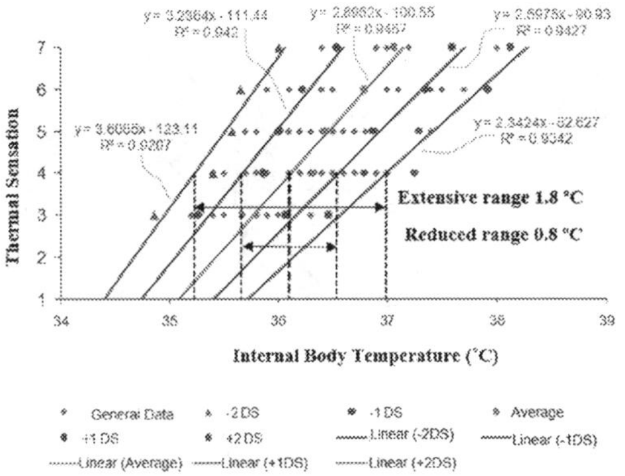

IBT NEUTRAL	IBTn-2DS	IBTn-1DS	IBTn-Average	IBTn+1DS	IBTn+2DS
	35.2	35.7	36.1	36.5	37.0

DS	Thermal Sensation	Scale	-2DS	-1DS	Average	+1DS	+2DS
0.53	Hot	7	36.0	36.5	37.1	37.6	38.1
0.57	Warm	6	35.7	36.2	36.8	37.4	37.9
0.43	Slightly warm	5	35.6	36.0	36.4	36.9	37.3
0.46	Neutral	4	35.4	35.9	36.3	36.8	37.3
0.40	Slightly cool	3	34.9	35.3	35.7	36.1	36.5

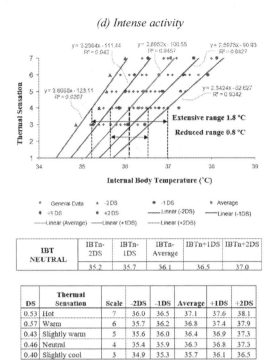

FIGURE 2.12 Neutral internal body temperature and thermal comfort ranges for: total observations, activities: passive, moderate, and intense.
Source: Elaborated by author.

2.5.2.2 THERMAL SENSATION BY MEAN SKIN TEMPERATURE (MST)

The thermal sensation by MST for the total amount of observations had a variation of 1.1 to 2.1C of mean values (Mean) of the perceived thermal sensation level, in reference to the value of thermal comfort. Regarding the estimated values of ± 1DS and ± 2DS per thermal sensation interval, a maximum variation of 10.8°C was obtained. It was observed that the regression lines were convergent as the thermal sensation of heat increased. The variation between the amplitude of the extended and the reduced range was 2.2°C. The difference of R2 for the five regressions presented a maximum variation of 0.18 (Figure 2.13(a)).

In the passive activity's case, the thermal sensation by MST, presented a variation of 1 to 2°C for the mean values (Mean) of the perceived thermal sensation level, in reference to the thermal comfort value. Regarding the estimated values of ± 1DS and ± 2DS per thermal sensation interval, a maximum variation of 10.8°C was obtained. It was observed that the regression lines were convergent as the heat sensations increased. The variation between the amplitude of the extended and the reduced range was 1.8°C. The difference of R2 for the five regressions presented a maximum variation of 0.18 (Figure 2.13(b)).

Meanwhile, for moderate activity's case, the thermal sensation by MST presented a variation of 1 to 1.9°C for the mean values (Mean) of the perceived thermal sensation level, in reference to the thermal comfort value. Regarding the estimated values of ± 1DS and ± 2DS per thermal sensation interval, a maximum variation of 10.2C was obtained (Figure 2.13(c)). It was observed that the regression lines merge at a point above the value 7 (very hot). The variation between the amplitude of the extended range and the reduced one was 1.7°C. The difference of R2 for the five regressions presented a maximum variation of 0.02.

The thermal sensation per MST for intense activity presented a variation of 1.1 to 2.3°C in reference to the thermal comfort value. Regarding the estimated values of ± 1DS and ± 2DS per thermal sensation interval, a maximum variation of 5.8°C was obtained (Figure 2.13(d)). It was observed that all the regression lines were convergent as the thermal sensation rose to "very hot." The variation between the amplitude of the extended and the reduced range was 4.5°C. The difference of R2 for the five regressions presented a maximum variation of 0.39.

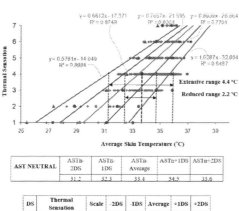

(a) Total observations

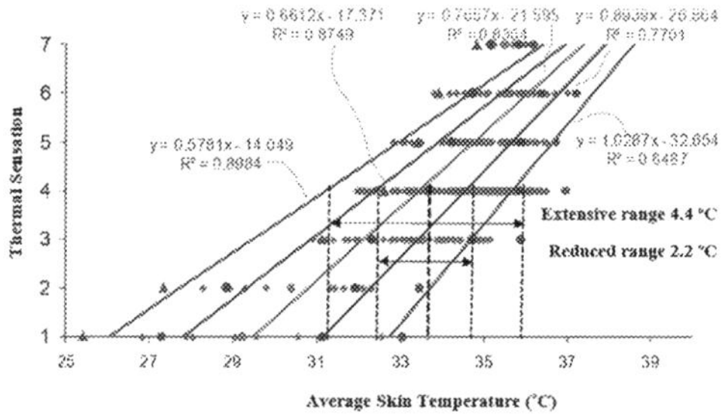

AST NEUTRAL	ASTn-2DS	ASTn-1DS	ASTn-Average	ASTn+1DS	ASTn+2DS
	31.2	32.3	33.4	34.5	35.6

DS	Thermal Sensation	Scale	-2DS	-1DS	Average	+1DS	+2DS
0.3	Hot	7	34.8	35.1	35.5	35.8	36.2
0.8	Warm	6	33.9	34.7	35.5	36.4	37.2
0.8	Slightly warm	5	33.4	34.2	35.1	35.9	36.7
1.1	Neutral	4	32.6	33.7	34.8	35.9	36.9
1.2	Slightly cool	3	31.1	32.3	33.5	34.7	35.9
1.5	Cool	2	27.4	28.9	30.4	31.9	33.4
1.9	Cold	1	25.4	27.3	29.2	31.1	33.0

FIGURE 2.13 *(Continued)*

(b) Passive activity

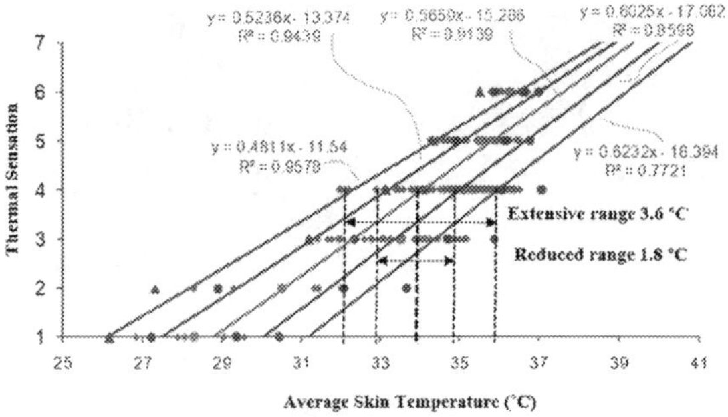

AST NEUTRAL	ASTn-2DS	ASTn-1DS	ASTn-Average	ASTn+1DS	ASTn+2DS
	32.3	33.2	34.1	35.0	35.9

DS	Thermal Sensation	Scale	-2DS	-1DS	Average	+1DS	+2DS
0.37	Warm	6	35.5	35.9	36.2	36.6	37.0
0.62	Slightly warm	5	34.3	34.9	35.5	36.2	36.8
0.99	Neutral	4	33.1	34.1	35.1	36.1	37.1
1.18	Slightly cool	3	31.2	32.3	33.5	34.7	35.9
1.59	Cool	2	27.3	28.9	30.5	32.1	33.7
1.07	Cold	1	26.2	27.2	28.3	29.4	30.4

(c) Moderate activity

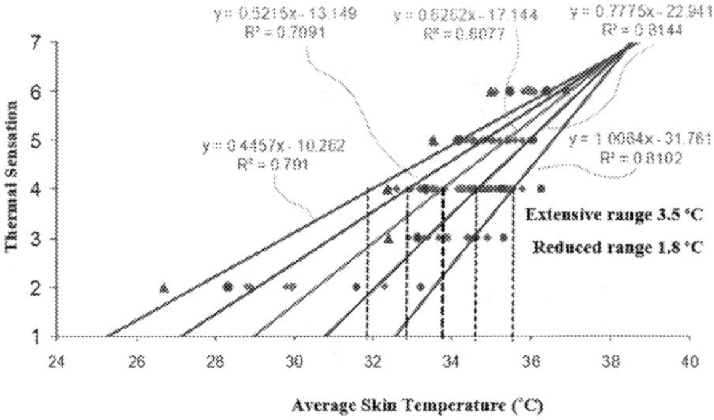

AST NEUTRAL	ASTn-2DS	ASTn-1DS	ASTn-Average	ASTn+1DS	ASTn+2DS
	32.0	32.9	33.8	34.7	35.5

DS	Thermal Sensation	Scale	-2DS	-1DS	Average	+1DS	+2DS
0.47	Warm	6	35.0	35.5	35.9	36.4	36.9
0.62	Slightly warm	5	33.5	34.2	34.8	35.4	36.0
0.97	Neutral	4	32.4	33.3	34.3	35.3	36.2
0.73	Slightly cool	3	32.4	33.1	33.9	34.6	35.3
1.63	Cool	2	26.7	28.3	30.0	31.6	33.2

FIGURE 2.13 *(Continued)*

(d) Intense activity

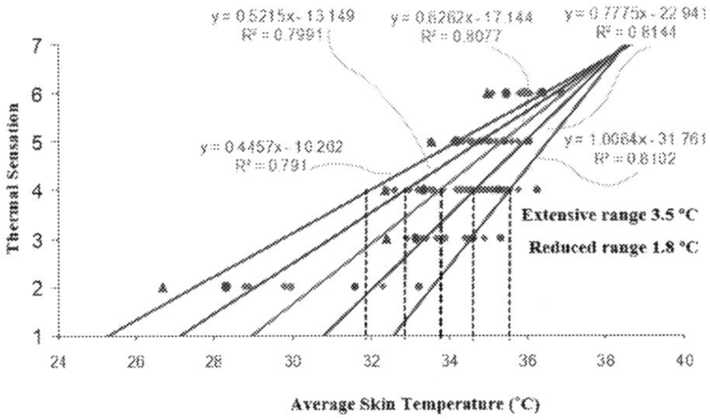

AST NEUTRAL	ASTn-2DS	ASTn-1DS	ASTn-Average	ASTn+1DS	ASTn+2DS
	32.0	32.9	33.8	34.7	35.5

DS	Thermal Sensation	Scale	-2DS	-1DS	Average	+1DS	+2DS
0.47	Warm	6	35.0	35.5	35.9	36.4	36.9
0.62	Slightly warm	5	33.5	34.2	34.8	35.4	36.0
0.97	Neutral	4	32.4	33.3	34.3	35.3	36.2
0.73	Slightly cool	3	32.4	33.1	33.9	34.6	35.3
1.63	Cool	2	26.7	28.3	30.0	31.6	33.2

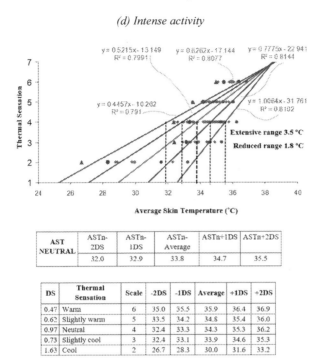

FIGURE 2.13 Mean neutral skin temperature and thermal comfort ranges for: total observations, activities: passive, moderate, and intense.
Source: Elaborated by author.

2.5.3 THERMAL SENSATION: ENVIRONMENTAL AND PHYSIOLOGICAL

In the thermal environmental temperature's (DBT) case, the neutral value for passive activity was 26.1°C, whereas for the total amount of observations the value was 5% lower than the passive one, the moderate and intense activity were 9% and 15% lower respectively in relation to passive activity (Table 2.9). The greatest variation in the temperatures recorded between passive and intense activity were observed as a result of the subjects' thermal adaptation effect in their metabolic processes. In keeping with the assumption of the balance thermodynamic exchange of Fanger's comfort equation, it was observed that those with a higher metabolic rate (intense activity) have a lower neutral temperature. The abovementioned demonstrates the importance of considering the experience and expectation effect on the perceived thermal sensation.

TABLE 2.9 Comparison of Neutral Values, Comfort Ranges, and Coefficient of Determination for DBT, IBT, and MST

Characteristic	Observation Total (°C)	Passive Activity (°C)	Moderate Activity (°C)	Intense Activity (°C)
	Transition Period			
	2192 Observations	1284 Observations	485 Observations	423 Observations
Neutral Temperature Based on Thermal Environment (DBTn)				
DBTn+2DS	31.4	32.9	29.1	27.6
DBTn+1DS	28.1	29.5	26.5	25.0
DBTn Mean	24.8	26.1	23.8	22.3
DBTn-1DS	21.5	22.7	21.1	19.7
DBTn-2DS	18.3	19.2	18.5	17.1
Extensive range	13.1	13.7	10.6	10.5
Reduced range	6.6	6.8	5.4	5.3
R^2(LRM)	0.9808	0.9804	0.9780	0.9580
Neutral Internal Body Temperature (IBTn)				
IBTn+2DS	37.3	37.1	37.1	37.0
IBTn+1DS	36.7	36.7	36.5	36.5
IBTn Mean	36.1	36.3	35.8	36.1
IBTn-1DS	35.4	35.9	35.1	35.7
IBTn-2DS	34.8	35.5	34.4	35.2
Extensive range	2.5	1.6	2.7	1.8
Reduced range	1.3	0.8	1.4	0.8
R^2(LRM)	0.9629	0.9613	0.9466	0.9467
Neutral Mean Skin Temperature (MSTn)				
MSTn+2DS	35.6	35.9	35.5	38.3
MSTn+1DS	34.5	35.0	34.7	35.0
MSTnMean	33.4	34.1	33.8	33.8
MSTn-1DS	32.3	33.2	32.9	32.6
MSTn-2DS	31.2	32.3	32.0	31.4
Extensive range	4.4	3.6	3.5	6.9
Reduced range	2.2	1.8	1.8	2.4
R^2(LRM)	0.8364	0.9139	0.8077	0.9361

Abbreviations: DBT*n*: Neutral dry bulb temperature; IBT*n*: Neutral internal body temperature; *MSTn*: Neutral mean skin temperature; *DS*: Standard deviation; R^2(LRM): Determining coefficient of the mean regression line.

Source: Elaborated by author.

The extensive range for thermal environmental temperature, in the passive activity, had a value of 13.7°C, whereas for the total amount of observations the value was 5% lower than the passive one, the moderate activity was 22% lower, and in the intense activity's case was 23% lower than passive. For the passive to intense activity levels, the values of extensive ranges were between 10.5 to 13.7°C, which was reasonable if it is considered that the study was carried out from 6:00 a.m. to 9:30 p.m. with a minimum of 13.7°C and maximum of 38.5°C. The widest range was revealed in subjects with passive activity, which is considered reasonable for a transition period with variations between morning and evening temperatures, as well as the variety of subjects and clothing within this activity level (Table 2.9).

The reduced range for neutral DBT had a value in the passive activity of 6.8°C, whereas for the total amount of observations the value was 3% lower than the passive one and in the moderate activity it was 21% lower. In the intense activity's case, it was 22% lower than the passive activity. For the passive to intense activity levels, the values of reduced ranges were between 5.3 and 6.8C; an acceptable variation if the study conditions mentioned above are considered. The broadest range occurred in subjects with passive activity, owing to what was mentioned in the previous extensive ranges (Table 2.9).

The determining coefficient for MST for passive activity had a value of 0.9804, whereas for the total amount of observations the value was 0.001% lower than the passive one. For moderate activity it was 0.002% lower, and in the intense activity's case, it was 2% lower than the passive activity.

As of the IBT, its neutral temperature due to passive activity was 36.3°C whereas for the total amount of observations the value was 5.50×10–3% lower than the passive one. The moderate activity was 1.3% lower than the passive activity and for the intense activity it was 5.50×10–3% lower than the passive one (Table 2.9).

It could be assumed that as the activity level increases, the IBT increases. However, similar conditions of IBT were observed for all the activity levels analyzed, with variations less than 1.3%. Therefore, in broad terms, and based on what was mentioned by Mondelo et al. (2001), regarding human thermoregulation, a general level of similar thermal adaptation can be considered in all cases. The abovementioned demonstrates the importance of considering the experience and expectation effect on the perceived thermal sensation; in addition, to the fact that thermal adaptation modified the functioning of human thermoregulation.

The extensive range for IBT had a value in passive activity of 1.6°C, whereas for the total amount of observations the value was 36% higher than the passive one and in moderate activity it was 40% higher. In the intense activity's case, it was 11% higher than passive one (Table 2.9). For the passive to intense activity levels, the values of extensive ranges were between 1.6 and 2.7°C, which was reasonable considering that the study was carried out between 6:00 a.m. and 9:30 p.m. with a minimum of 13.7°C and maximum of 38.5°C. The widest range occurred in subjects with moderate activity, due to the variety of types of activities within this range, the level of clothing, and the temperature conditions for observations during this level of activity.

During the passive activity, the reduced range for IBT had a value of 0.8°C, whereas for the total amount of observations the value was 38% higher than the passive activity and in moderate activity it was 43% higher than the passive one. In the intense activity's case, it was the same as passive (Table 2.9). For the passive to intense activity levels, the values of reduced ranges were between 0.8 and 1.4C, an acceptable variation considering the study conditions mentioned above. The widest range was presented in subjects with moderate activity, due to what was mentioned in the previous extensive ranges.

During the passive activity, the determining coefficient for IBT had a value of 0.9613, whereas for the total amount of observations the value was 1.66×10^{-3}% higher than the passive one and in moderate activity it was 1.5% lower. In the intense activity's case, it was 1.5% lower than the passive activity (Table 2.9). It is worth mentioning that the determining coefficient is reduced from the passive activity to moderate activity, while the moderate and intense are similar.

In the MST's case, the neutral value for passive activity was 34.1°C, whereas for the total amount of observations the value was 2% lower than the passive one. The moderate and intense activities were 8.79×10^{-3}% lower than passive activity (Table 2.9). A slight amount of variation was observed in the recorded temperatures, attributable to the subjects' thermal adaptation effect. It could be assumed that as the level of activity increases, the MST increases; however, similar conditions were observed, with variations less than 2%. Therefore, in broad terms, and based on what was mentioned by Mondelo et al. (2001), regarding human thermoregulation, a similar general level of thermal adaptation can be considered in all cases. The abovementioned demonstrates the importance of considering the experience and expectation effect on the perceived thermal sensation.

During the passive activity, the extensive range for MST had a value of 3.6°C, whereas for the total amount of observations the value was 18% higher than the passive one and in the moderate activity it was 3% lower.

In the intense activity's case, it was 48% higher than the passive activity (Table 2.9). For the passive to intense activity levels, the extensive range values were between 3.6 and 6.9°C, which was reasonable, considering that the study was carried out from 6:00 a.m. to 9:30 p.m. with a minimum of 13.7°C and maximum of 38.5°C. The widest range was presented in subjects with intense activity; possibly due to the convective effect of wind speeds generated by the movement of the study subjects, the level of clothing and the temperature conditions for the observations at this level of exercise.

During the passive activity, the reduced range for MST had a value of 1.8°C, whereas for the total amount of observations the value was 18% greater than passive one and in moderate activity it was the same as passive. In the intense activity's case, it was 25% greater than passive activity (Table 2.9). For the passive to intense activity levels, the reduced range values were between 1.8 and 2.4°C, an acceptable variation considering the study conditions mentioned above. The widest range was exhibited in subjects with intense activity, due to what was mentioned in the previous extensive ranges.

For passive activity, the determining coefficient for MST had a value of 0.9139, whereas for the total amount of observations the value was 8% lower than the passive one and in the moderate activity it was 12% lower. In the intense activity's case, it was 2% greater than the passive activity (Table 2.9).

Regarding the neutral values (DBTn, IBTn, MSTn) it can be observed, through their mean value, that in the thermal environment effect's (DBTn) case, it has 24.8°C for the total amount of observations, 26°C for the passive activity 23.8°C for moderate activity, and 22.3°C for intense activity. This decrease is due to the subjects' metabolic level, as well as the thermodynamic interaction with the thermal environment, which allows a better adaptation to a lower temperature the intense activity's case. In the case of IBTn, it can be observed that there are values close to 36°C, due to the body's metabolic stabilization requirements, which indicates a good physiological response of the surveyed subjects to the environmental conditions. In the MST's case, its mean neutral value per activity level is approximately 2° different from the IBTn (Table 2.9).

But if the IBT values are compared with the DBT, a difference of 11.3°C is observed for all of the observations, 10.2°C for passive activity, 12°C for moderate activity and 13.8°C for intense activity. These differences and increases clearly establish the subjects' levels of adaptation that are due to the discipline of the exercise that they carry out every day (Table 2.9).

In MST's case in reference with DBT, there is a difference of 11.9°C for the total of observations, 8°C for passive activity, 10C for moderate activity, and 11.5°C for intense activity. In this case, it is observed that the skin remains in different conditions than the body's core. However, it reflects the adaptation process prior to the metabolic conditions established by the IBT. Nevertheless, the subjects' adaptation process is presented (Table 2.9). The same conditions of the neutral values (DBT, IBT, MST) are presented for both the extended and the reduced range, in the same proportion for each case.

2.6 CONCLUSIONS

Urban thermal environments contain physical and biological factors that directly affect the users carrying out activities in their environment. These factors significantly influence the users' physiological and psychological health, consequently, the relationship between availability and access to these urban environments considerably influences society as a motivator for physical activities that help reduce the risk of disease.

Thermal comfort in urban spaces determines the density and periodicity of its use. It is necessary to understand the causal relations that make a recreational or cohabiting space to be underused due to environmental conditions. This is the reason that it is necessary to analyze the spaces' bioclimate, and its modifications, based on its conditioning factors and the direct impact on the perceived thermal sensation.

Regarding the thermal adaptation based on the neutral temperature and physiological temperature and the mean of the neutral values (DBTn, IBTn, MSTn), it can be observed that the thermal environment effect (DBTn), for the total amount of observations, shows a greater degree of adaptation in subjects who carry out an intense activity and a similar adaptation of subjects in moderate activity. This is due to the regular practice of this level of activity and to the thermodynamic interaction with the urban thermal environment, which demonstrates that as you exercise (intense and moderate activity) you not only have better health, but you also improve in terms of adaptation to the climate.

The IBTn maintains values close to what is considered normal with a range of 35.8 to 36.3°C, which demonstrates that the mean temperature required for the core of the body is in accordance. However, for the MSTn the ranges are between 33.4 to 34.1°C, the difference between the IBTn and MSTn is congruent if MSTn is considered as an area of the body exposed to urban thermal ambient conditions, and the IBTn measurement does not present this feature. On the other hand, it is also shown that the same thermodynamic process marks a differential of 2.2 to 2.4°C between the external surface of the body and its nucleus.

It is observed with the analysis of DBTn compared with IBTn and MSTn, that in the body nucleus (IBTn) the subject's adaptation level is reflected and also that the skin (MSTn) is maintained in conditions different from the body nucleus; nevertheless, it reflects the process of adaptation prior to the metabolic conditions established by the IBT.

This investigation provides pertinent information regarding the use and assessment of the human thermal environment and its effect on the user's thermal sensation and adaptation. The results obtained can be used for the design of outdoor spaces or as reference values in light of risky conditions, due to morbidity, in conditions of critical thermal environment. They can also be used as risk indicators of the climate effect on human conditions.

ACKNOWLEDGMENTS

I would like to thank the architect Ana Teresa Soberanes Lopez for all the support provided in the preparation of this document, to the support staff for the organization and carrying out of the fieldwork used for this investigation. I would like to thank the Recreation Center *Juventud 2000*, Universidad Autonoma de Baja California, and Universidad de Colima, for all their support and LucaBoJi and ZuliBoJi, PhD, for their time in helping carry out this investigation.

KEYWORDS

- **dry bulb temperature**
- **equivalent physiological temperature**
- **internal body temperature**
- **mean skin temperature**
- **means by thermal sensation interval**
- **standard deviation**
- **World Health Organization**

REFERENCES

Aerts, R., Honnay, O., & Nieuwenhuyse, A., (2018). Biodiversity and human health: mechanisms and evidence of the positive health effects of diversity in nature and green spaces. *British Medical Bulletin, 127*(1), 5–22.

American Society of Heating, Refrigerating, and Air conditioning Engineers, (2017). *ANSI/ASHRAE 55-2017: Thermal Environmental Conditions for Human Occupancy*. Atlanta: Autor.

Auliciems, A., & Szokolay, S., (1997). *Thermal Comfort*. Notes of passive and low energy architecture international. No. 3. Brisbane: PLEA – University of Queensland.

Aziz, N., & Said, I., (2015). Outdoor environments as children's play spaces: Playground affordences. En: Evans, B., et al., (eds.), *Play*, and *Recreation, Health, and Wellbeing, Geographies of Children and Young People* (Vol. 9). Springer, Singapore.

Bojórquez-Morales, G., Gómez-Azpeitia, G., García-Cueto, R., García-Gómez, C., Luna-León, A., & Romero-Moreno, R., (2012). Neutral temperature in outdoors for warm and cold periods for extreme warm dry climate. *Proceedings of the 7th Windsor Conference: The Changing Context of Comfort in an Unpredictable World Cumberland Lodge*, Windsor, UK, London: Network for Comfort and Energy Use in Buildings. http://nceub.org.uk (accessed on 21 December 2021).

Brager, G., & Dear De, R., (1998). Thermal adaptation in the built environment: a literature review. *Energy and Buildings, 27*, 83–96.

Chang, P., (2020). *Effects of the Built and Social Features of Urban Greenways on the Outdoor Activity of Older Adults*, 204.

Chee, A., Jordan, H., & Horsley, J., (2015). *Value of Urban Green Spaces in Promoting Healthy Living and Wellbeing: Prospects for Planning, 8*, 131–137. Recuperado de. doi: 10.2147/RMHP.S61654

Cooper, C., & Francis, C., (1998). *People Places: Design Guidelines for Urban Open Spaces*. New York-Toronto: Jhon Wiley Sons, Inc.

Del, C. N. U., & González, J. N., (2020). Bioclimatic, environmental measurements and use of urban spaces: Comparative evaluation in the Plaza de Chamberí, Madrid. *Revistarquis, 9*(1), 1–26.

Díaz, R., Castro, A., & Aranda, P., (2013). *Mortality Due to Excessive Natural Heat in Northwestern Mexico: Social Conditions Associated with this Cause of Death* (Vol. 26, No. 52). Northern Border, Mexico. ISSN: 2594-0260.

Elnabawi, M., & Hamza, N., (2019). Behavioral perspectives of outdoor thermal comfort in urban areas: a critical review. *Atmosphere, 11*, 51. doi: 10.3390/atmos11010051.

Fanger, O., (1972). *Thermal Comfort.* New York: McGraw-Hill.

Fanger, P. O., (1986). Thermal environment – human requirements. *The Environmentalist, 6*(4), 275–278. Springer Netherlands.

Galarraga, P., Vives, M., Manzano-Cabrera, D., Urda, L., Brito, & Gea-Caballero, V., (2018). The incorporation of community health in the planning and transformation of the urban environment. *SESPAS 2018 Report, 32*(S1), 74–81. Recovered from: doi: https://doi.org/10.1016/j.gaceta.2018.08.001

Ghysais, G., (2018). *Collective Facilities and Construction of Urban Identity to Produce Competitiveness in the Municipality of Sincelejo*. Master's thesis in Urbanism and Territorial Development. School of Architecture, Urbanism, and Design, Barranquilla, Colombia.

Gómez-Azpeitia, G., Ruiz, R., Bojórquez, G., & Romero, R., (2007). *Monitoring of Thermal Comfort Conditions Product 3*. (National Housing Fund Commission. Thermal comfort and energy saving project in affordable housing in Mexico, regions with hot, dry, and humid climates. CONAFOVI. 2004-01-20). Colima, Colima.

Höppe, P., (2002). Different aspects of assessing indoor and outdoor thermal comfort. *Energy and Building, 34*, 661–665.

Humphreys, M., & Nicol, F., (2002). The validity of ISO-PMV for predicting comfort votes in every-day thermal environments. *Energy and Buildings, 34,* 667–684.

Hwang, R. L., & Lin, T. P., (2007). Thermal comfort requirements for occupants of semi-outdoor and outdoor environments in hot-humid regions. *Architectural Science Review, 50*(4), 357–364.

Inbar, O., Morris, N., Epstein, Y., &, Gass, G., (2004). Comparison of thermoregulatory responses to exercise in dry heat among prepubertal boys, young adults and older males. *Exp. Physiology, 89*(6), 691–700.

Inter-ministerial Commission on Climate Change (CICC), (2009). *"Special Climate Change Program 2009-2012."* Official Gazette of the Federation, Mexico DF.

International Organization for Standardization, (2005). *ISO 7730:2005 (E) Ergonomics of the Thermal Environment – Analytical Determination and Interpretation of Thermal Comfort using Calculation of the PMV and PPD Indices and Local Thermal Comfort Criteria*. Ginebra: Autor.

International Organization for Standardization, (2005). *ISO 8996:2005 (E) Ergonomics of the Thermal Environment – Determination of Metabolic Heat Production*. Ginebra: Autor.

International Organization for Standardization. 7726, (1998). *Ergonomics of the Thermal Environment – Instruments for Measuring Physical Quantities*. Ginebra: International organizations for Standardizations.

International Organization for Standardization. ISO 10551, (2019). *Ergonomics of Thermal Environment – Assessment of the Influence of the Thermal Environment Using Subjective Judgement Scales* (p. 52). International Organization for Standardization, Ginebra.

Jennings, V., Larson, L., & Yun, J., (2015). Advancing Sustainability through urban Green Space: cultural ecosystem services, equity, and social determinants of health. *En: International Journal of Environmental research and Public Helath, 13*(2), 196. Recuperado de: doi: https://doi.org/10.3390/ijerph13020196

Karis, C., Magalí, C., & Ferrero, R., (2019). Environmental indicators and urban management. Relations between urban ecosystem services and sustainability. In: *Urban Notebook* (Vol. 27, No. 27, pp. 9–30). Space, Culture, Society.

Kim, S., & Jeong, W., (2002). Influence of illumination on autonomic thermoregulation and choice of clothing. *Int. Journal Biometeoroly, 46*, 141–144.

Lin, T. P., (2009). Thermal perception, adaptation, and attendance in a public square in hot and humid regions. *Building and Environment, 44*(10), 2017–2026.

Lin, T. P., De Dear, R., Hwang, R. L., (2011). Effect of thermal adaptation on seasonal outdoor thermal comfort. *International Journal of Climatology, 31*(2), 302–312.

Luna, A., (2008). *Design and Evaluation of Energetically Sustainable Housing*. Doctoral thesis. Institute of Engineering, Autonomous University of Baja California, Mexicali.

Mondelo, P., Gregori, E., Comas, S., Castejón, E., & Bartolomé, E., (2001). *Ergonomics 2: Comfort and Heat Stress* (3rd edn.). Barcelona: Universitat Politècnica Catalunya.

Nicol, F., (1993). *Thermal Comfort "A Handbook for Field Studies Toward an Adaptive Model."* London, University of East London.

Nikolopoulou, M., (2004). *Designing Open Space in the Urban Environment: A Bioclimatic Approach.* Attiki: Center for renewable energy sources.

Nikolopoulou, M., & Steemers, K., (2003). Thermal comfort and psychological adaptation as a guide for designing urban spaces. *Energy and Buildings, 35*, 95–101.

Oliveira, S., & Andrade, H., (2007). An initial assessment of the bioclimatic comfort in an outdoor public space in Lisbon. *International Journal of Biometeorology, 52*(1), 69–84.

Pickup, J., & De Dear, R., (2000). An outdoor thermal comfort index (OUT_SET*): Part I – the model and its assumptions. In: De Dear, R. J., Kalma, J. D., Oke, T. R., & Auliciems, A., (eds.), *Biometeorology*, and *Urban Climatology at the Turn of the Millennium* (pp. 279–283). WCASP 50: WMO/TD No.1026. WMO: Geneva.

Potter, J., & De Dear, R., (2000). Field study to calibrate an outdoor thermal comfort index. In: De Dear, R. J., Kalma, J. D., Oke, T. R., & Auliciems, A., (eds.), *Biometeorology*, and *Urban Climatology at the Turn of the Millennium* (pp. 315–320). WCASP 50: WMO/TD No.1026. WMO: Geneva.

Quijije, C., & Castro, J., (2020). Interaction of public spaces in urban dynamics. *Scientific Journal Domain of the Sciences, 6*(2), 539–553. Recovered from: doi: http://dx.doi.org/10.23857/dc.v6i2.1183.

Rhoades, R., & Tanner, G., (1997). *Fisiología Médica*. Barcelona: Masson.

Southon, G., Jorgensen, A., Dunett, N., Hoyle, H., &Evans, K., (2018). Perceived spaces-richness in urban green spaces: Cues, accuracy, and wellbeing impacts. En: *Landscape and Urban Planning, 172*, 1–10.

Spagnolo, J., & deDear, R., (2003). A field study of thermal comfort in outdoor and semi-outdoor environments in subtropical Sydney Australia. *Building and Environment, 38*(5), 721–738.

Urias-Barrera, H., (2019). *Thermal Comfort in Outdoor Public Spaces for Intense Sports Activities: In Extreme Hot Dry Weather*. Mimeo. Doctoral thesis. Autonomous University of Baja California.

Vidal, R., (2004). *Paths Trodden, Paths to Conquer: Towards the Bioclimatization of Outdoor Public Spaces*. Scientific and Technological Contributions, Year 1, Number 1. https://www.academia.edu/38214641/2004_Senderos_andados_caminos_por_conquistar_pdf (accessed on 21 December 2021).

Wood, L., Hooper, P., Foster, S., & Bull, F., (2017). *Public Green Spaces and Positive Mental Health – Investigating the Relationship Between Access, Quantity, and Types of Parks and Mental Wellbeing, 48*, 63–71.

World Health Organization, (2018). *Mental Health: Strengthening our Response*. In World Health Organization. Retrieved from: https://www.who.int/es/news-room/fact-sheets/detail/mental-health-strengthening-our-response (accessed on 21 December 2021).

CHAPTER 3

The Use of Alternative Materials with Disinfectant Characteristics in the Face of the Pandemic of the SARS-CoV-2

RUBÉN SALVADOR ROUX GUTIÉRREZ

Institute of Higher Studies of Tamaulipas, A. C. Av. Dr. Burton E. Grossman 501 Pte, Zip Code: 89605 Altamira, Tamaulipas, Mexico, E-mail: ruben.roux@iest.edu.mx

3.1 INTRODUCTION

3.1.1 THE COVID-19 CAUSED BY SARS-COV-2

The disease caused by a new coronavirus (COVID-19) that developed in Wuhan, a city in Hubei province, in China, at the end of 2019, registered an accelerated worldwide spread with an exponential increase in the number of cases of infections and deaths (Pan American Health Organization/World Health Organization, 2020). At the time of writing this document, the disease had 24,776,988 infected and 837,979 dead people worldwide (Google, 2020). This pandemic has radically changed the lifestyle of all the people in the world. Since there is currently no treatment endorsed by the World Health Organization (WHO), nor an approved vaccine, the best recommendation for the population has been confinement in their homes and take protective measures such as washing their hands frequently with soap and water, wearing a mask, learn to sneeze or cough using a tissue, clean, and disinfect, using spray products or wet wipes, and practice social distancing (HealthyChildren.org, 2020).

Architecture for Health and Well-Being: A Sustainable Approach. Maria Eugenia Molar Orozco, PhD (Ed.)

In the case of confinement, adequate sanitation of the surfaces has been recommended by using suitable chemicals that can kill the virus, which based on studies has been determined that it can last up to 72 hours on plastic and stainless-steel surfaces, less than 4 hours on copper surfaces, as well as less than 24 hours on cardboard surfaces (WHO, 2020). In disinfection processes, it is recommended to check the spectrum of effectiveness as well as the risks to human health due to its toxicity (KAHRS, 1995).

The arrival of the disease to the American continent was announced on January 30, 2020. The WHO Director-General declared the COVID-19 outbreak as a Public Health Emergency of International Concern (PHEIC) in accordance with the International Health Regulations (2005). The first case detected in the Americas Region was confirmed in the United States on January 20, 2020, followed by Brazil, which registered the first case on February 26, 2020. Since that date, COVID-19 has spread to the 54 countries and territories of the Americas Region (Pan American Health Organization/World Health Organization, 2020).

In Mexico, the first contagion was announced on February 27, 2020:

> *The first confirmed case was presented in Mexico City, and it was a Mexican who had traveled to Italy and had mild symptoms; a few hours later, another case was confirmed in the state of Sinaloa, and a third case, again, in Mexico City. The first death from this disease in the country occurred on March 18, 2020 (Anonymous, 2020).*

On August 29, 2020, Mexico is in eighth place in relation to the number of infections with 585,738, and in third place worldwide according to the number of deaths with 63,146 (Salud, 2020). One of the problems that have been reflected due to the pandemic in Mexico has been the fall in the economy, which in the second quarter of the year is 18.7% of the gross domestic product (GDP) (Informador, 2020). Why is it important to make this reference? Because of the fact that to maintain the health measures indicated by the WHO, it is necessary to be able to invest in the purchase of all the required inputs and at this time the economy of a large percentage of the Mexican population is not favorable, and if to this we add that although they are not expensive, their application must be daily, it is not always possible to do this.

Therefore, the possibility of using a traditional material such as Calcium Hydroxide ($Ca(OH)_2$) is raised. Its use as a disinfectant is intrinsic in the culture and customs of society. Throughout history it has had multiple applications since its antibacterial and fungicidal effect has been acquired empirically (Ruiz, Ponce, and Alvarado, 1995), and it has already been used as a disinfectant in other cases such as avian flu, the brucellosis, the equine encephalitis virus, among other animal diseases (European Lime Association, 2009). In other cases, it has also been used to combat Cholera (Vibrio Cholerae 01), in aqueous solutions, as well as against other microorganisms such as: *Streptococcus mufans*, *Peptostreptococcus anaerobius*, *Porphyromonas gingivalis*, and *Fusobac-ferium nucleatum* in this case as a paste in Dentistry since 1929 (Muñoz Ruiz, Collazo Ponce, and Alvarado, 1995). Calcium Hydroxide is therefore considered a high impact disinfectant product that can be used to sterilize objects, tools, and spaces that may be exposed to a type of contamination. It has been proven that its property as a disinfectant is due to its high alkalinity; this characteristic does not support the development of potentially infectious organisms for humans (Cerón, 2016).

Calcium hydroxide can be used in construction in various ways, one as a plaster, another as a paint and as a cementitious agent in some masonry materials, which would allow to have surfaces (walls and ceilings) with disinfectant characteristics that are susceptible to SARS-CoV-2 remaining on these surfaces.

3.1.2 TYPES OF DISINFECTANTS

The World Health Organization defines disinfectants as:

> *A disinfectant is a chemical agent that destroys or inhibits the growth of pathogenic microorganisms in vegetative or non-sporulated phase (WHO, 2004).*

Below it will be mentioned the types of disinfectants recommended to combat the SARS-CoV-2 produced by COVID-19, as well as its main characteristics and restrictions to avoid health problems (see Table 3.1).

Although it is true that disinfectants currently serve as the main ally to keep the surfaces and spaces in which man develops free of bacteria, fungi, and viruses, the reality is that, although they are effective, there are determining factors that decrease their percentage of utility due to their mismanagement. Since cleaning is necessary prior to the application of this, in addition to that the effectiveness depends on other factors such as the type of organism, its way of multiplying, and its resistance in the environment, as well as to the chemical substances that are applied (Shuña and Lozano, 2016).

TABLE 3.1 List N: Products with Emerging Viral Pathogens AND Human Coronavirus Claims for Use Against SARS-CoV-2

EPA se Registration Number	Active Ingredient(s)	Product Name	Company	Follow the Disinfection and Preparation Instructions for the Following Viruses	Contact Time (in minutes)	Formulation Type	Surface Types	Use Sites	Declaration of Emerging Viral Pathogen	Date Added to List N
1839-216	Quaternary ammonium	SC-NDC-64	Stepan Company	Human coronavirus	5	Dilutable	Hard, Nonporous (HN); required	Healthcare; Institutional	No	4/30/2020
777-139	Citric acid	T-bone	Reckitt Benckiser LLC	Human coronavirus	5	Wipe	Hard, nonporous (HN); Food contact post-rinse required (FCR)	Healthcare; institutional; residential	No	4/30/2020
91176-2	1,2-Hexanediol	PELS 422	The Filla Company LLC	Human coronavirus	10	Ready-to-use	Hard, nonporous (HN)	Healthcare; institutional; residential	No	4/23/2020
954-11	Quaternary ammonium	Barbicide	King Research Inc.	Human coronavirus	10	Dilutable	Hard nonporous (HN)	Healthcare; institutional	No	4/02/2020
777-128	Quaternary ammonium	Lysol® Laundry Sanitizer	Reckitt Benckiser LLC	Human coronavirus	5	Dilutable (laundry presoak only)	Porous (P) (laundry presoak only)	Residential	No	3/19/2020
10324-59	Quaternary ammonium	Maquat 64	Mason Chemical Company	Human coronavirus	10	Dilutable	Hard, Nonporous (HN)	Healthcare; institutional; residential	No	3/19/2020
10324-108	Quaternary ammonium	Maquat 256-MN	Mason Chemical Company	Human coronavirus	10	Dilutable	Hard, nonporous (HN); food contact post-rinse required (FCR)	Healthcare; institutional; residential	No	3/13/2020

TABLE 3.1 *(Continued)*

EPA se Registration Number	Active Ingredient(s)	Product Name	Company	Follow the Disinfection and Preparation Instructions for the Following Viruses	Contact Time (in minutes)	Formulation Type	Surface Types	Use Sites	Declaration of Emerging Viral Pathogen	Date Added to List N
10324-112	Quaternary ammonium	Maquat 128-MN	Mason Chemical Company	Human coronavirus	10	Dilutable	Hard, nonporous (HN); food contact post-rinse required (FCR)	Healthcare; institutional; residential	No	3/13/2020
10324-113	Quaternary ammonium	Maquat 64-MN	Mason Chemical Company	Human coronavirus	10	Dilutable	Hard, nonporous (HN); food contact post-rinse required (FCR)	Healthcare; institutional; residential	No	3/13/2020
10324-59	Quaternary ammonium	Maquat 64	Mason Chemical Company	Human coronavirus	10	Dilutable	Hard, Nonporous (HN)	Healthcare; Institutional; Residential	No	3/19/2020
10324-115	Quaternary ammonium	Maquat 750-M	Mason Chemical Company	Human coronavirus	10	Dilutable	Hard, nonporous (HN)	Healthcare; institutional; residential	No	3/13/2020
10324-117	Quaternary ammonium	Maquat 710-M	Mason Chemical Company	Human coronavirus	10	Dilutable	Hard, nonporous (HN); food contact post-rinse required (FCR); porous (P) (laundry presoak only)	Healthcare; institutional; residential	No	3/13/2020
10324-140	Quaternary ammonium	Maquat MQ2525M-CPV	Mason Chemical Company	Human coronavirus	10	Dilutable	Hard, nonporous (HN); food contact post-rinse required (FCR)	Healthcare; institutional;	No	3/13/2020

TABLE 3.1 *(Continued)*

EPA se Registration Number	Active Ingredient(s)	Product Name	Company	Follow the Disinfection and Preparation Instructions for the Following Viruses	Contact Time (in minutes)	Formulation Type	Surface Types	Use Sites	Declaration of Emerging Viral Pathogen	Date Added to List N
10324-141	Quaternary ammonium	Maquat 256-NHQ	Mason Chemical Company	Coronavirus humano	10	Dilutable	Hard, nonporous (HN)	Healthcare; institutional; residential	No	3/13/2020
10324-142	Quaternary ammonium	Maquat MQ2525M-14	Mason Chemical Company	Human coronavirus	10	Dilutable	Hard, nonporous (HN); food contact post-rinse required (FCR)	Healthcare; institutional; residential	No	3/13/2020
10324-154	Quaternary ammonium	Maquat 64-NHQ	Mason Chemical Company	Human coronavirus	10	Dilutable	Hard, nonporous (HN)	Healthcare; institutional; residential	No	3/13/2020
10324-155	Quaternary ammonium	Maquat 128-NHQ	Mason Chemical Company	Human coronavirus	10	Dilutable	Hard, nonporous (HN)	Healthcare; institutional; residential	No	3/13/2020
10324-156	Quaternary ammonium	Maquat 512-NHQ	Mason Chemical Company	Human coronavirus	10	Dilutable	Hard, nonporous (HN)	Healthcare; institutional; residential	No	3/13/2020
10324-157	Quaternary ammonium	Maquat 32-NHQ	Mason Chemical Company	Human coronavirus	10	Dilutable	Hard, nonporous (HN)	Healthcare; institutional; residential	No	3/13/2020
10324-164	Quaternary ammonium	Maquat 256-PD	Mason Chemical Company	Human coronavirus	10	Dilutable	Hard, nonporous (HN)	Healthcare; institutional; residential	No	3/13/2020
10324-166	Quaternary ammonium	Maquat 32	Mason Chemical Company	Human coronavirus	10	Dilutable	Hard, nonporous (HN); food contact post-rinse required (FCR)	Healthcare; institutional; residential	No	3/13/2020

TABLE 3.1 *(Continued)*

EPA se Registration Number	Active Ingredient(s)	Product Name	Company	Follow the Disinfection and Preparation Instructions for the Following Viruses	Contact Time (in minutes)	Formulation Type	Surface Types	Use Sites	Declaration of Emerging Viral Pathogen	Date Added to List N
10324-167	Quaternary ammonium	Maquat 32-PD	Mason Chemical Company	Human coronavirus	10	Dilutable	Hard, nonporous (HN)	Healthcare; institutional; residential	No	3/13/2020
10324-177	Quaternary ammonium	Maquat 705-M	Mason Chemical Company	Human coronavirus	10	Dilutable	Hard, nonporous (HN); food contact post-rinse required (FCR); porous (P) (laundry presoak only)	Healthcare; institutional; residential	No	3/13/2020
10324-194	Quaternary ammonium	Maquat	Mason Chemical Company	Human coronavirus	10	Dilutable	Hard, nonporous (HN)	Healthcare; institutional; residential	No	3/13/2020
10324-198	Quaternary ammonium	Maquat 702.5-M	Mason Chemical Company	Human coronavirus	10	Dilutable	Hard, nonporous (HN); porous (P) (laundry presoak only)	Healthcare; institutional; residential	No	3/13/2020
10324-214	Hydrogen peroxide; peroxyacetic acid	Maguard 5626	Mason Chemical Company	Human coronavirus	10	Dilutable	Hard, nonporous (HN)	Healthcare; institutional; residential	No	3/13/2020
10324-230	Hydrogen peroxide; peroxyacetic acid (peracetic acid)	Maguard 1522	Mason Chemical Company	Human coronavirus	1	Dilutable	Hard, nonporous (HN)	Healthcare; institutional; residential	No	03/13/2020
10324-57	Quaternary ammonium	Maquat 42	Mason Chemical Company	Human coronavirus	10	Dilutable	Hard, nonporous (HN)	Healthcare; institutional; residential	No	03/13/2020

TABLE 3.1 *(Continued)*

EPA se Registration Number	Active Ingredient(s)	Product Name	Company	Follow the Disinfection and Preparation Instructions for the Following Viruses	Contact Time (in minutes)	Formulation Type	Surface Types	Use Sites	Declaration of Emerging Viral Pathogen	Date Added to List N
10324-58	Quaternary ammonium	Maquat 128	Mason Chemical Company	Human coronavirus	10	Dilutable	Hard, nonporous (HN)	Healthcare; institutional; residential	No	03/13/2020
10324-63	Quaternary ammonium	Maquat 10	Mason Chemical Company	Human coronavirus	10	Dilutable	Hard, nonporous (HN)	Healthcare; institutional; residential	No	03/13/2020
10324-71	Quaternary ammonium	Maquat 280	Mason Chemical Company	Human coronavirus	10	Dilutable	Hard, nonporous (HN)	Healthcare; institutional; residential	No	03/13/2020
10324-72	Quaternary ammonium	Maquat 615-HD	Mason Chemical Company	Human coronavirus	10	Dilutable	Hard, nonporous (HN); food contact post-rinse required (FCR)	Healthcare; institutional; residential	No	03/13/2020
10324-80	Quaternary ammonium	Maquat 5.5-M	Mason Chemical Company	Human coronavirus	10	Dilutable	Hard, nonporous (HN)	Healthcare; institutional; residential	No	03/13/2020
10324-81	Quaternary ammonium	Maquat 7.5-M	Mason Chemical Company	Norovirus; feline calicivirus	10	Dilutable	Hard, nonporous (HN); food contact post-rinse required (FCR); porous (P) (laundry presoak only)	Healthcare; institutional; residential	No	06/04/2020
10324-85	Quaternary ammonium	Maquat 86-M	Mason Chemical Company	Hepatitis A virus; Porcine rotavirus	10	Ready-to-use	Hard, nonporous (HN); food contact post-rinse required (FCR)	Healthcare; institutional; residential	No	06/08/2020
10324-93	Quaternary ammonium	Maquat 64-PD	Mason Chemical Company	Human coronavirus	10	Dilutable	Hard, nonporous (HN)	Healthcare; institutional; residential	–	03/13/2020

TABLE 3.1 *(Continued)*

EPA se Registration Number	Active Ingredient(s)	Product Name	Company	Follow the Disinfection and Preparation Instructions for the Following Viruses	Contact Time (in minutes)	Formulation Type	Surface Types	Use Sites	Declaration of Emerging Viral Pathogen	Date Added to List N
10324-94	Quaternary ammonium	Maquat 20-M	Mason Chemical Company	Human coronavirus	10	Dilutable	Hard, nonporous (HN); food contact post-rinse required (FCR)	Healthcare; institutional; residential	No	03/13/2020
10324-96	Quaternary ammonium	Maquat 50-DS	Mason Chemical Company	Human coronavirus	10	Dilutable	Hard, nonporous (HN); food contact post-rinse required (FCR)	Healthcare; institutional; residential	No	03/13/2020
10324-99	Quaternary ammonium	Maquat 10-PD	Mason Chemical Company	Human coronavirus	10	Dilutable	Hard, nonporous (HN)	Healthcare; institutional; residential	No	03/13/2020
10492-4	Quaternary ammonium; isopropanol (isopropyl alcohol)	Discide ultra disinfecting towelettes	Palmero Healthcare LLC	Adenovirus Type 2	1	Wipe	Hard, nonporous (HN)	Healthcare; institutional; residential	No	06/11/2020
10492-5	Quaternary ammonium; isopropanol (isopropyl alcohol)	Discide ultra disinfecting spray	Palmero Healthcare LLC	Human coronavirus	0.5 (30 seconds)	Ready-to-use	Hard, nonporous (HN)	Healthcare; institutional; residential	No	03/13/2020
11346-4	Quaternary ammonium	Clorox QS	The Clorox Company	Human coronavirus	2	Ready-to-use	Hard, nonporous (HN); food contact post-rinse required (FCR)	Healthcare; residential	No	03/13/2020
1672-65	Sodium hypochlorite	Austin A-1 ultra-disinfecting bleach	James Austin Company	Hepatitis A virus	10	Dilutable	Hard, nonporous (HN)	Healthcare; institutional; residential	No	053/21/2020

TABLE 3.1 *(Continued)*

EPA se Registration Number	Active Ingredient(s)	Product Name	Company	Follow the Disinfection and Preparation Instructions for the Following Viruses	Contact Time (in minutes)	Formulation Type	Surface Types	Use Sites	Declaration of Emerging Viral Pathogen	Date Added to List N
1672-67	Sodium hypochlorite	Austin's A-1 Concentrated Bleach 8.25%	James Austin Company	Human coronavirus	5	Dilutable	Hard, nonporous (HN)	Healthcare; institutional; residential	No	03/13/2020
1677-204	Octanoic acid	65 disinfecting heavy-duty acid bathroom cleaner	Ecolab Inc	Human coronavirus	2	Dilutable	Hard, nonporous (HN)	Healthcare; institutional	No	03/13/2020
1677-241	Sodium hypochlorite	Hydris	Ecolab Inc	Human coronavirus	5	Ready-to-use	Hard, nonporous (HN)	Healthcare; institutional	No	03/13/2020
1839-167	Quaternary ammonium	BTC 885 neutral disinfectant cleaner-256	Stepan Company	Rotavirus	10	Dilutable	Hard, nonporous (HN)	Healthcare; institutional; residential	No	05/28/2020
1839-168	Quaternary ammonium	BTC 885 NDC-32	Stepan Company	Rotavirus	10	Dilutable	Hard, nonporous (HN); food contact post-rinse required (FCR)	Healthcare; institutional; residential	No	05/28/2020
1839-169	Quaternary ammonium	BTC 885 neutral disinfectant cleaner-64	Stepan Company	Rotavirus	10	Dilutable	Hard nonporous (HN); food contact post-rinse required (FCR)	Healthcare; institutional; residential	No	06/08/2020
1839-176	Quaternary ammonium	Liquid-pak neutral disinfectant cleaner	Stepan Company	Human coronavirus	10	Dilutable	Hard, nonporous (HN); food contact post-rinse required (FCR)	Healthcare; institutional; residential	No	03/13/2020
1839-190	Quaternary ammonium	Stepan disinfectant wipe	Stepan Company	Human coronavirus	10	Wipe	Hard, nonporous (HN)	Healthcare; institutional; residential	No	03/13/2020
1839-214	Quaternary ammonium	SC-NDC-256	Stepan Company	Human coronavirus	5	Dilutable	Hard, nonporous (HN); food contact post-rinse required (FCR)	Healthcare; institutional; residential	No	03/13/2020

TABLE 3.1 *(Continued)*

EPA se Registration Number	Active Ingredient(s)	Product Name	Company	Follow the Disinfection and Preparation Instructions for the Following Viruses	Contact Time (in minutes)	Formulation Type	Surface Types	Use Sites	Declaration of Emerging Viral Pathogen	Date Added to List N
1839-78	Quaternary ammonium	NP 3.2 detergent/ disinfectant	Stepan Company	Human coronavirus	10	Dilutable	Hard, nonporous (HN); food contact post-rinse required (FCR)	Healthcare; institutional; residential	No	03/13/2020
1839-79	Quaternary ammonium	NP 4.5 detergent/ disinfectant	Stepan Company	Human coronavirus	10	Dilutable	Hard, nonporous (HN); Food contact post-rinse required (FCR)	Healthcare; institutional; residential	No	03/13/2020
1839-81	Quaternary ammonium	NP 9.0 detergent/ disinfectant	Stepan Company	Human coronavirus	10	Dilutable	Hard, nonporous (HN); food contact post-rinse required (FCR)	Healthcare; institutional; residential	No	03/13/2020
1839-94	Quaternary ammonium	NP 3.2 (D&F) Detergent/ disinfectant	Stepan Company	Norovirus	10	Dilutable	Hard, nonporous (HN)	Healthcare; institutional; residential	No	06/17/2020
3862-191	Quaternary ammonium	Assure	ABC Compounding Co Inc	Human coronavirus	10	Dilutable	Hard, nonporous (HN)	Healthcare; institutional; residential	No	03/13/2020
4091-23	Sodium hypochlorite; sodium carbonate	Mold armor formula 400	W.M. Barr and Company Inc	Human coronavirus	0.5 (30 seconds)	Ready-to-use	Hard, nonporous (HN)	Institutional residential;	No	03/13/2020
42964-17	Quaternary ammonium; ethanol (ethyl alcohol)	Asepticare	Airkem professional products	Human coronavirus	2	Ready-to-use	Hard, nonporous (HN)	Healthcare; institutional; residential	No	03/13/2020
46781-6	Quaternary ammonium; isopropanol (isopropyl alcohol)	Cavicide	Metrex Research	Human coronavirus	2	Ready-o-use	Hard, nonporous (HN)	Healthcare; institutional; residential	No	03/13/2020

TABLE 3.1 *(Continued)*

EPA se Registration Number	Active Ingredient(s)	Product Name	Company	Follow the Disinfection and Preparation Instructions for the Following Viruses	Contact Time (in minutes)	Formulation Type	Surface Types	Use Sites	Declaration of Emerging Viral Pathogen	Date Added to List N
4822-548	Triethylene glycol; quaternary ammonium	Scrubbing Bubbles® Multi-Purpose Disinfectant	S.C. Johnson and Son Inc	Rotavirus	5	Pressurized liquid	Hard, nonporous (HN)	Residential	No	06/04/2020
4822-606	L-Lactic acid	Fangio	S.C. Johnson and Son Inc	Human coronavirus	10	Ready-to-use	Hard, nonporous (HN)	Institutional; residential	No	03/13/2020
4822-607	Quaternary ammonium	Lauda	S.C. Johnson and Son Inc	Human coronavirus	5	Ready-to-use	Hard, nonporous (HN); food contact post-rinse required (FCR)	Institutional; residential	No	03/13/2020
4822-608	L-Lactic acid	Gurney	S.C. Johnson and Son Inc	Human coronavirus	5	Ready-to-use	Hard nonporous (HN)	Institutional; residential	No	03/13/2020
4822-609	Quaternary ammonium	Stewart	S.C. Johnson and Son Inc	Human coronavirus	3	Ready-to-use	Hard, nonporous (HN)	Institutional; residential	No	03/13/2020
54289-4	Peroxyacetic acid (peracetic acid)	Peraclean 15 (peroxyacetic acid solution)	Evonik Corporation	Human coronavirus	3	Dilutable	Hard, nonporous (HN)	Healthcare; institutional	No	03/13/2020
56392-10	Sodium hypochlorite	Caltech Swat 200 9B	Clorox professional products company	Human coronavirus	2	Ready-to-use	Hard, nonporous (HN); food contact post-rinse required (FCR)	Healthcare; institutional	No	03/13/2020
5813-103	Sodium hypochlorite	CGB 3	The Clorox Company	Human coronavirus	5	Dilutable	Hard, nonporous (HN)	Healthcare; institutional; residential	No	03/13/2020
5813-104	Sodium hypochlorite	CGB4	The Clorox Company	Human coronavirus	5	Dilutable	Hard, nonporous (HN)	Healthcare; institutional; residential	No	03/13/2020

TABLE 3.1 *(Continued)*

EPA se Registration Number	Active Ingredient(s)	Product Name	Company	Follow the Disinfection and Preparation Instructions for the Following Viruses	Contact Time (in minutes)	Formulation Type	Surface Types	Use Sites	Declaration of Emerging Viral Pathogen	Date Added to List N
5813-73	Quaternary ammonium	Clorox Everest	The Clorox Company	Human coronavirus	0.5 (30 seconds)	Ready to use	Hard, nonporous (HN)	Institutional; residential	No	03/13/2020
5813-86	Glycolic acid	CBW	The Clorox Company	Human coronavirus	10	Impregnated materials	Hard, nonporous (HN)	Residential	No	03/13/2020
5813-98	Sodium hypochlorite	Lite	The Clorox Company	Human coronavirus	1	Ready to use	Hard, nonporous (HN); food contact post-rinse required (FCR)	Institutional; residential	No	03/13/2020
5813-99	Sodium hypochlorite	Wave	The Clorox Company	Human coronavirus	1	Wipe	Hard, nonporous (HN)	Institutional; residential	No	03/13/2020
61178-1	Quaternary ammonium	D-125	Microgen Inc	Human coronavirus	10	Dilutable	Hard, nonporous (HN)	Healthcare; institutional; residential	No	03/13/2020
61178-5	Quaternary ammonium	CCX-151	Microgen Inc	Human Coronavirus	10	Dilutable	Hard, nonporous (HN)	Healthcare; institutional; residential	No	03/13/2020
6198-4	Quaternary ammonium	Q. A. Concentrated	Nationals Chemicals Inc.	Human coronavirus	10	Dilutable	Hard, nonporous (HN); food contact post-rinse required (FCR)	Healthcare; institutional; residential	No	03/13/2020
62472-2	Quaternary ammonium	Kennelsol HC	Alpha Tech Pet Inc.	Human coronavirus	10	Dilutable	Hard, nonporous (HN)	Institutional; residential	No	03/13/2020
67619-10	Quaternary ammonium	CPPC Everest	Clorox Professional Products Company	Human coronavirus	10	Dilutable	Hard, nonporous (HN); food contact post-rinse required (FCR)	Healthcare; institutional; residential	No	03/13/2020

TABLE 3.1 *(Continued)*

EPA se Registration Number	Active Ingredient(s)	Product Name	Company	Follow the Disinfection and Preparation Instructions for the Following Viruses	Contact Time (in minutes)	Formulation Type	Surface Types	Use Sites	Declaration of Emerging Viral Pathogen	Date Added to List N
67619-11	Sodium hypochlorite	CPPC Shower	Clorox professional products company	Human coronavirus	1	Ready to use	Hard, nonporous (HN); food contact post-rinse required (FCR)	Healthcare; institutional; residential	No	03/13/2020
67619-13	Sodium hypochlorite	CPPC Storm	Clorox Professional Products Company	Human coronavirus	1	Ready to use	Hard, nonporous (HN); food contact post-rinse required (FCR)	Healthcare; institutional; residential	No	03/13/2020
67619-27	Sodium hypochlorite	Buster	Clorox Professional Products Company	Human coronavirus	5	Ready-to-use	Hard, nonporous (HN)	Healthcare; institutional; residential	No	03/13/2020
67619-28	Sodium hypochlorite	Milo	Clorox Professional Products Company	Human coronavirus	5	Dilutable	Hard, nonporous (HN)	Healthcare; institutional; residential	No	03/13/2020
67619-8	Sodium hypochlorite	CPPC Ultra Bleach 2	Clorox Professional Products Company	Human coronavirus	5	Dilutable	Hard, nonporous (HN)	Healthcare; institutional; residential	No	03/13/2020
6836-336	Quaternary ammonium	Lonza disinfectant wipes plus	Lonza LLC	Human coronavirus	4	Wipe	Hard, nonporous (HN)	Healthcare; institutional; residential	No	03/13/2020
6836-381	Quaternary ammonium	Lonzagard R-82G	Lonza LLC	Human coronavirus	1	Dilutable	Hard, nonporous (HN)	Healthcare; institutional; residential	No	03/13/2020
6836-382	Quaternary ammonium	Nugen low streak disinfectant wipes	Lonza LLC	Human coronavirus	4	Wipe	Hard, nonporous (HN)	Healthcare; institutional; residential	No	03/13/2020

TABLE 3.1 *(Continued)*

EPA se Registration Number	Active Ingredient(s)	Product Name	Company	Follow the Disinfection and Preparation Instructions for the Following Viruses	Contact Time (in minutes)	Formulation Type	Surface Types	Use Sites	Declaration of Emerging Viral Pathogen	Date Added to List N
70590-2	Sodium hypochlorite	Bleach-rite disinfecting spray with bleach	Current Technologies Inc	Human coronavirus	1	Ready-to-use	Hard, nonporous (HN)	Healthcare; institutional; residential	No	03/13/2020
70627-15	Quaternary ammonium	Warrior	Diversey Inc	Human coronavirus	10	Dilutable	Hard, nonporous (HN); food contact post-rinse required (FCR)	Healthcare; institutional	No	03/13/2020
70627-2	Quaternary ammonium	Disinfectant D.C 100	Diversey Inc.	Human coronavirus	2	Ready-to-use	Hard, nonporous (HN); food contact post-rinse required (FCR)	Healthcare; institutional	No	03/13/2020
70627-23	Quaternary ammonium	Virex™ II/64	Diversey Inc	Human coronavirus	10	Dilutable	Hard, nonporous (HN); food contact post-rinse required (FCR)	Healthcare; institutional	No	03/13/2020
70627-6	Phenolic	Phenolic disinfectant HG	Diversey Inc	Human coronavirus	10	Dilutable	Hard, nonporous (HN); food contact post-rinse required (FCR)	Healthcare; institutional	No	03/13/2020
70627-62	Hydrogen peroxide	Phato 1:64 disinfectant cleaner	Diversey Inc.	Human coronavirus	5	Dilutable	Hard, nonporous (HN)	Healthcare; institutional	No	03/13/2020
70627-63	Quaternary ammonium	512 sanitizers	Diversey Inc	Human coronavirus	10	Dilutable	Hard, nonporous (HN)	Healthcare; institutional	No	03/13/2020
70627-75	Sodium hypochlorite	Avert sporicidal disinfectant cleaner wipes	Diversey Inc.	Human coronavirus	1	Wipe	Hard, nonporous (HN)	Healthcare; institutional	No	03/13/2020

TABLE 3.1 *(Continued)*

EPA se Registration Number	Active Ingredient(s)	Product Name	Company	Follow the Disinfection and Preparation Instructions for the Following Viruses	Contact Time (in minutes)	Formulation Type	Surface Types	Use Sites	Declaration of Emerging Viral Pathogen	Date Added to List N
70627-78	Hydrogen peroxide	Suretouch	Diversey Inc	Human coronavirus	5	Ready-to-use	Hard, nonporous (HN)	Healthcare; institutional	No	03/13/2020
72977-5	Silver ion; citric acid	Sdc3a	ETI H_2O Inc	Human coronavirus	1	Ready-to-use	Hard, nonporous (HN)	Healthcare; institutional; residential	No	03/13/2020
74559-6	Hydrogen peroxide	Oxy-res (concentrate)	Virox Technologies Inc.	Human coronavirus	5	Dilutable	Hard, nonporous (HN)	Healthcare; institutional; residential	No	03/13/2020
74559-8	Hydrogen peroxide	Accel 5 RTU	Virox Technologies Inc	Human coronavirus	5	Ready-to-use	Hard, nonporous (HN)	Healthcare; institutional; residential	No	03/13/2020
777-136	Ethanol (ethyl alcohol)	Lysol® Neutra Air® 2	Reckitt Benckiser LLC	Human coronavirus	0.5 (30 seconds)	Ready-to-use	Hard, nonporous (HN); food contact post-rinse required (FCR)	Healthcare; institutional; residential	No	03/13/2020
74986-4	Sodium chlorite	Selectrocide 2L500	Selective Micro Technologies LLC	Human coronavirus	10	Dilutable	Hard, nonporous (HN)	Healthcare; institutional	No	03/13/2020
74986-5	Sodium chlorite	Selectrocide 5g	Selective Micro Technologies LLC	Human coronavirus	10	Solid	Hard, nonporous (HN)	Healthcare; institutional	No	03/13/2020
777-130	Quaternary ammonium	Caterpillar	Reckitt Benckiser LLC	Human coronavirus	2.5 (2 minutes and 30 seconds)	Wipe	Hard, nonporous (HN)	Healthcare; institutional; residential	No	03/13/2020

TABLE 3.1 *(Continued)*

EPA se Registration Number	Active Ingredient(s)	Product Name	Company	Follow the Disinfection and Preparation Instructions for the Following Viruses	Contact Time (in minutes)	Formulation Type	Surface Types	Use Sites	Declaration of Emerging Viral Pathogen	Date Added to List N
8383-14	Hydrogen peroxide Peroxyacetic acid (peracetic acid)	PeridoxRTU (brand) one-step germicidal wipes	Contec Inc	Human coronavirus	0.5 (30 seconds)	Wipe	Hard, nonporous (HN)	Healthcare; institutional	No	03/13/2020
777-66	Quaternary ammonium	Lysol® brand all-purpose cleaner	Reckitt Benckiser LLC	Human coronavirus	2	Ready-to-use	Hard, nonporous (HN)	Healthcare; institutional; residential	No	03/13/2020
777-82	Quaternary ammonium	Lysol® brand deodorizing disinfectant cleaner	Reckitt Benckiser LLC	Human coronavirus	10	Dilutable	Hard, nonporous (HN)	Institutional; residential	No	03/13/2020
777-91	Quaternary ammonium	Lysol® Kitchen pro antibacterial cleaner	Reckitt Benckiser LLC	Human coronavirus	2	Ready-to-use	Hard, nonporous (HN); food contact post-rinse required (FCR)	Healthcare; institutional;	No	03/13/2020
8383-7	Phenolic	Sporicidin (Brand) disinfectant towelettes	Contec Inc	Human coronavirus	5	Wipe	Hard, nonporous (HN)	Healthcare; institutional; residential	No	03/13/2020
85343-1	Quaternary ammonium	Teccare Control	Talley Environmental Care Limited	Human coronavirus	10	Dilutable	Hard, nonporous (HN)	Healthcare; institutional; residential	No	03/13/2020
88494-1	Quaternary ammonium ethanol (ethyl alcohol)	Wedge disinfectant	North American Infection Control LTD	Human coronavirus	1	Dilutable	Hard, nonporous (HN)	Healthcare; institutional; residential	No	03/13/2020
89896-2	Hypochlorous acid	Cleansmart	Simple Science Limited	Human coronavirus	10	Ready-to-use	Hard, nonporous (HN); food contact no-rinse (FCNR)	Healthcare; institutional; residential	No	03/13/2020

TABLE 3.1 *(Continued)*

EPA se Registration Number	Active Ingredient(s)	Product Name	Company	Follow the Disinfection and Preparation Instructions for the Following Viruses	Contact Time (in minutes)	Formulation Type	Surface Types	Use Sites	Declaration of Emerging Viral Pathogen	Date Added to List N
89900-1	Hydrogen peroxide	Nathan 2	S.C. Johnson Professional	Human coronavirus	5	Ready-to-use	Hard, nonporous (HN)	Healthcare; institutional; residential	No	03/13/2020
90287-1	Quaternary ammonium	Maquat 25.6-PDX	VI-JON Inc	Human coronavirus	10	Dilutable	Hard, nonporous (HN)	Healthcare; institutional; residential	No	03/13/2020
9402-14	Hydrogen peroxide Ammonium carbonate; Ammonium bicarbonate	Hitman Spray	Kimberly-Clark Global Sales LLC	Human coronavirus	5	Ready-to-use	Hard, nonporous (HN)	Institutional; residential	No	03/13/2020
89900-1	Hydrogen peroxide	Nathan 2	S.C. Johnson Professional	Human coronavirus	5	RTU[2]	Hard, nonporous (HN)	Healthcare; institutional; residential	No	3/13/2020
90287-1	Quaternary ammonium	Maquat 25.6-PDX	VI-JON Inc	Human coronavirus	10	Dilutable	Hard, nonporous (HN)	Healthcare; institutional; residential	No	3/13/2020
9402-14	Hydrogen peroxide; ammonium carbonate; ammonium bicarbonate	Hitman spray	Kimberly-Clark Global Sales LLC	Human coronavirus	5	RTU	Hard, nonporous (HN)	Institutional; residential	No	3/13/2020
9402-15	Hydrogen peroxide; ammonium carbonate; ammonium bicarbonate	Victor spray	Kimberly-Clark Global Sales LLC	Human coronavirus	5	Pressurized liquid	Hard, nonporous (HN)	Healthcare; institutional; residential	No	3/13/2020

TABLE 3.1 *(Continued)*

EPA se Registration Number	Active Ingredient(s)	Product Name	Company	Follow the Disinfection and Preparation Instructions for the Following Viruses	Contact Time (in minutes)	Formulation Type	Surface Types	Use Sites	Declaration of Emerging Viral Pathogen	Date Added to List N
9402-17	Hydrogen peroxide; Ammonium carbonate; ammonium bicarbonate	Hitman wipe	Kimberly-Clark Global Sales LLC	Human coronavirus	6	Wipe	Hard, nonporous (HN)	Institutional; residential	No	3/13/2020
9480-5	Quaternary ammonium	Sani-cloth germicidal disposable cloth	Professional Disposables International Inc.	Human coronavirus	3	Wipe	Hard, nonporous (HN); food contact post-rinse required (FCR)	Healthcare; institutional; residential	No	3/13/2020

Source: U.S. Environmental Protection Agency (2020).

As it can be seen in Table 3.1, none of the disinfectant substances works on porous surfaces. This characteristic decreases their disinfection effectiveness on the surfaces of architectural components such as walls and ceilings, whose manufacturing materials have a porous surface (concrete, masonry, mortar) and its maximum duration of effectiveness is 10 minutes, so the use of calcium carbonate applied as calcium hydroxide as paint can improve these conditions.

3.2 HISTORY OF THE USE OF LIME

It is not possible to know exactly when humans first discovered lime. However, there is the possibility that ancient civilizations used limestone to protect their stoves, thus producing the first burnt lime in history, due to the heating of the rocks with the produced fire. The same one that later became hydrated with the rain to form calcium hydroxide, which would end up reacting with the ashes of the fire pit and the sand, forming what could be considered as the first mortar (Lhoist, 2018).

The information on the background of the use of lime does not establish precisely when it was the first time that humans used lime as a binder. The first traces of its use date back to the Cretan culture of the Mediterranean more than 3,000 years ago, also there is evidence that lime mortars were used on the Chinese wall (Soluciones Prácticas, 1994).

There are traces of lime foundations in Turkey, denoting that it was already in use 14,000 years ago. On the other hand, in older evidence dating back to almost 16,000 years ago, frescoes using natural pigments of iron oxide-containing calcium or limestone were found applied on the wet stone walls in the Lascaux caves, in France (Lhoist, 2018).

In its use as cladding, the oldest example can be found in Anatolia, in the settlement of Çatalhöyük, dating from approximately 6600–5650 BC, (Figure 3.1) where there were wooden pillars which were covered with a mixture of lime and red pigment. In the work "Earliest Civilization of the Near East," James Mellaart describes the interior of the houses in the following way: "each of the houses (...) was equipped with wooden pillars covered with a mixture of lime painted red and the floor was treated in the same way. The walls were covered with beautiful schematic drawings of animals (...)" (Mellaart, 1965).

There is also information on the use of lime as paint and stucco in the region that the Arabs dominated in Spain, specifically in the Alhambra in

Granada and in the Mosque of Córdoba. In Iraq there are stuccoes made in Warta dating back to 6,000, as well as in Greece, which were used as a mortar in the construction of raw masonry walls. This technique of using lime was continued to be used in the Middle Ages, although at this time it concentrated on the use of generally aerial local limes, lime-based paint (also called liming) became popular in Europe during this period due to the disinfecting properties of the material, as well as its breathable and fire-retardant capacity. In a decree promulgated by King John of England in 1212, all the houses in and around the Támesis were ordered to be plastered and whitewashed (Cerón and C., 2016) It was used mainly outside the houses, stables, barracks, and guardhouses. The use of it in cities extended until the early 1900s, and in rural areas until the middle of the 20th century (Levano, Navarro, and Rosell, 2018).

FIGURE 3.1 Reconstruction of the interior of a house in Çatalhöyük.
Source: Natalia Franco drawing based on the reconstruction made by Cox (undated). 3D Artist. Obtained from: Çatalhöyük Shrine of the Hunters Looking South.

It is until the Industrial Revolution that it is discovered that limes can have hydraulic properties, due to the origin of the clays it contains; in England clay limes are discovered and the calcination process is improved, which means that there are two slopes, one that leads to the discovery of Portland cement as it is known today and the other that perfects hydraulic limes by means of the French engineer Louis Joseph Vicat, who develops in-depth studies of the difference between aerial and hydraulic lime.

But lime has not only been used in the construction area. There are data that indicate that in the year 160 BC. Lime was used for agricultural purposes in the manufacture of ovens (Usedo Valles, 2015). Another important antecedent is that in *"The Ten Books on Architecture"* by Marco Vitruvio Pollion. The use of lime is indicated and in what proportions it is you must mix with other types of aggregates to obtain a suitable mortar. According to Vitrubio's recommendation, the thickness of the stucco should be 5 to 8 centimeters, and then it should be decorated with fresco paintings, which were pigments diluted with lime water. Later, on some occasions, these paintings were dry retouched with tempera (Cerón, 2016).

In 1639 Fray Laurencio de San Nicolás also described the proportions of the arid aggregates to make masonry base mortars and plastered walls (Usedo Valles, 2015).

In Mesoamerica, there is a history of the use of lime in the Mexica, Maya, Totonaca, Zapotec, and Mixtec cultures. It can be indicated that the use of lime by these cultures was one of the main elements of their development. A clear example is the city of Teotihuacan which had 22 square kilometers of urban extension and where more than 100,000 people coexisted. The archaeological studies carried out at the site indicate that there were 2,200 housing complexes where an average of 45 people lived per group and it is in this great Mexica city where the use of lime has been discovered due to the production of pavements and coatings, prevailing over other types of materials. It is also known that this culture did not have health problems with epidemics. It is important to mention that in the highland area where the city of Teotihuacan is located, it is not an area with large deposits of limestone as in other regions of Mexico (Barba Pingarrón and Villaseñor Alonso, 2013).

Lime as a construction material in these cultures occurs with fewer variants since the stones of the buildings they made were generally used as mortar. An important example is what was done by the Mayans; thanks to the use of this material, it is inferred that they could make their vaults (see Figure 3.2) and crests of their great architectural works, unlike the area of the Mexican highlands where the mud mortar was used to join the mentioned stones (Barba Pingarrón and Villaseñor Alonso, 2013).

Lime was also used to construct buildings that will store natural products, such is the case of the so-called *"Chultunes."* These deposits were used in the Mayan culture to store water since they were made by digging in the limestone to later cover the walls with lime. This technology continued to be used until colonial times when European technologies brought by the

Spaniards were adapted, thus taking advantage of the use of rainwater, which was collected on the roofs and conducted to these tanks where the water was kept in good condition due to the lime coatings on their walls. Later, as well as lime was used in these tanks, other techniques used to store perishable materials such as corn kernel in warehouses or cuexcomates have also been found where lime milk was used on the walls interiors to prevent harmful insects from staying inside.

FIGURE 3.2 Mayan Vault in Uxmal, Yucatán.
Source: Drawing based on original photo in the book "Lime History: Properties and Use," author Luis Barba Pingarrón and Isabel Villaseñor Alonso on page 21.

In food processes, there is also a history of the use of lime, such as the nixtamalization of corn (see Figure 3.3). This technique allowed Mesoamerican cultures to use lime for the rest of the planet. Since the use of lime in this process with corn generated the tortilla (Barba Pingarrón and Villaseñor Alonso, 2013).

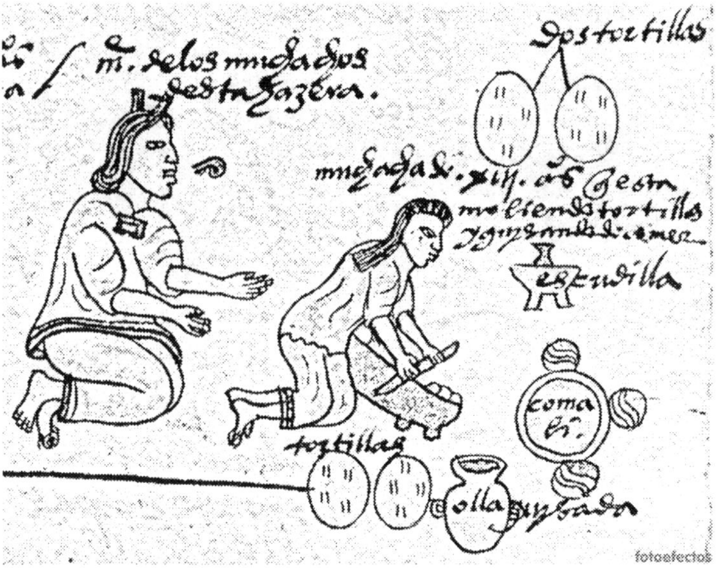

FIGURE 3.3 Pre-Hispanic nixtamalization process.
Source: Own elaboration; drawing based on the original drawing in the article of (García Gallegos, 2019).

Luis Fernando Guerrero Baca mentions in the chapter *"La cal y los sistemas constructivos"* that ancient cultures discovered the use of lime in a fortuitous way when they saw that the rocks when calcined, then moistened and later dried, acquired a binding characteristic which allowed the use of these elements in place of the mud that had been used to join the masonry elements. This undoubtedly resulted in a technological advance that changed the sense of architecture.

They discovered that lime was more resistant to meteorological phenomena than the other materials used up to that moment, however, its use was not very extensive since the raw materials to obtain it are not everywhere and its manufacturing processes are more complex than other materials, so it can be inferred that due to this reason it was not used in the same proportion as the masonry.

As it was previously mentioned, lime has been used throughout history by various civilizations as a building material and used as a finish in buildings.

In Europe, peasants used lime to sanitize the areas where the animals were located, as well as to fertilize their land. Once the epidemics of typhus, plague, and yellow fever arrived, the use of lime paint in outdoors and indoors of houses became a trend to prevent the proliferation of infections, due that its antiseptic and antibacterial effect was discovered (Aguilar, 2020).

Among the most important properties of lime is that it is fungicidal and biocidal, since it contains a high percentage of alkalinity, which prevents both bacteria and microorganisms from adhering to surfaces covered with this material, greatly reducing respiratory diseases and allergies from mites, mold, and bacteria in indoor architectural areas (Aguilar, 2020).

History tells us the importance of the use of lime in human life both in its food and its sanitation, conservation, and construction activities. At this moment, lime takes on a fundamental importance due to these characteristics that make it a material with multiple possibilities in the sanitation of human habitat.

3.2.1 LIME ORIGIN

"The limestone is a rock composed of at least 50% calcium carbonate ($CaCO_3$), with variable percentages of impurities, in its broadest interpretation, the term includes any calcareous material containing calcium carbonate such as marble, chalk, travertine, coral, and marl" (Guerrero Hernández, 2001), it is considered to be a stratified calcareous rock, with a high content of calcite, which gives rise to the product known as lime in a calcination process, if the rock in its weathering has a large content of iron carbonate, it will give as a product an iron oxide, however, most limestone rocks in their weathering process have little iron and greater quantities of silica and alumina. In nature, there is a variety of limestone, such as: biothermal limestone that contains a large number of skeletons of marine organisms from reef areas; the biostromics are composed similarly to the previous ones but occupy larger areas and in strata with variable thickness and between each stratum a layer of clay is located; Bituminous, their characteristic color is black since they are composed of kerosene and asphalt, which under a distillation process can extract fuel.

Commercial limestones are classified by their magnesium content in: Those that contain at least 5% are denominated as magnesian limestone, which contains between 30% and 40% of calcium carbonate, they are denominated as dolomitic limestones, which are composed of dolomite material, which is a double magnesium calcium carbonate ($CaCO_3MgCO_3$), that contains a 46% of magnesium carbonate, this type of rocks produced lime rich in calcium, magnesian lime and dolomitic lime (see Table 3.2 and Figure 3.4) (Guerrero Hernández, 2001).

TABLE 3.2 Chemical and Mineralogical Composition

Percentage of Dolomite (Mineral)	Percentage of Calcite	Rock Type
0–10	90–100	Limestone
10–50	50–90	Magnesian limestone or dolomitic limestone
50–90	10–50	Calcareous dolomite
90–100	0–10	Dolomite (dolomitic rock)

Source: Own elaboration based on website data: https://geologiaweb.com/rocas-sedimentarias/caliza/.

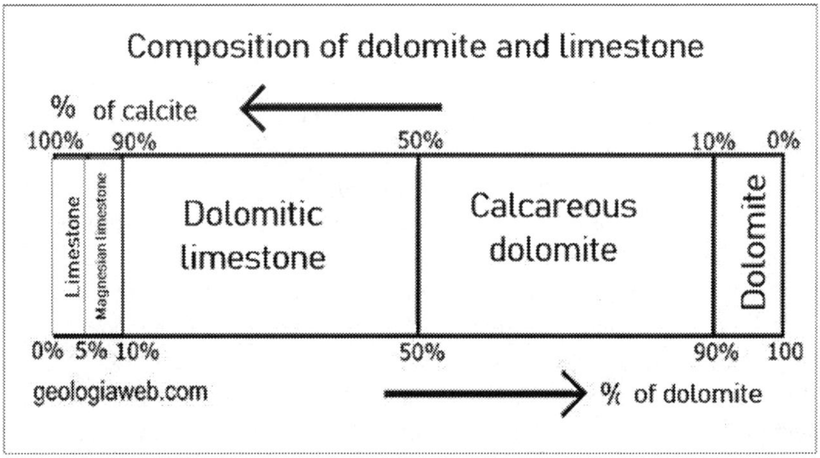

FIGURE 3.4 Dolomite and limestone composition.
Source: Own elaboration.

The states in Mexico which own deposits of limestone are: Campeche, Nuevo León, Quintana Roo, Yucatán, Hidalgo, Chiapas, San Luis Potosí, Veracruz, Guerrero, and Tamaulipas, the production of these states is mostly used in the construction industry, the chemical, agrochemical industry and in the manufacture of glass.

10% of this volume is limestone rock with calcite mineral ($CaCO_3$) specially used for produced lime (see Figure 3.5) and the subproducts derived from this (General Direction of Mining Development, 2018).

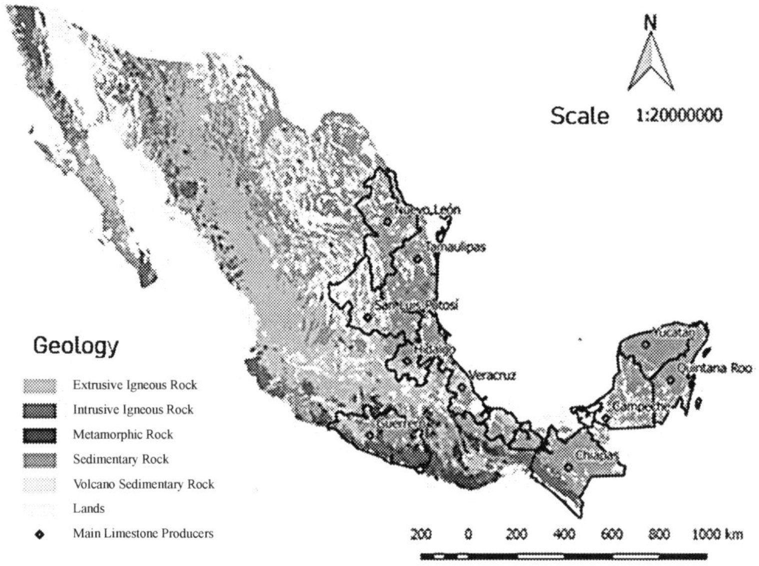

FIGURE 3.5 Main limestone producing states in Mexico.
Source: Own elaboration based on drawing in General Direction of Mining Development (2018), p. 7.

3.2.2 LIME CYCLE

The transformation of limestone to lime is a dynamic and cyclical process, this one begins with the extraction of limestone ($CaCO_3$) in the deposits, to later be crushed and thus carry out the calcination process in a furnace at a temperature between 900 a 1000C, to produced aerial limes, carbon dioxide (CO_2) is released at this stage, modifying its chemical composition leaving a calcium monoxide (CaCO) or quicklime; to do the lime slaking, it is submerged in water to achieve the slaking, you must mix 1 kg lime and 2.5 l of water and there is a chemical change again, producing calcium hydroxide ($Ca(OH)_2$) or slaked lime (see Figure 3.6 and Table 3.3), once lime has been applied as a mortar or paint, a chemical change begins again by absorbing CO_2 from the atmosphere, to turn into stone again (National Network of Teachers of Traditional Construction, 2018).

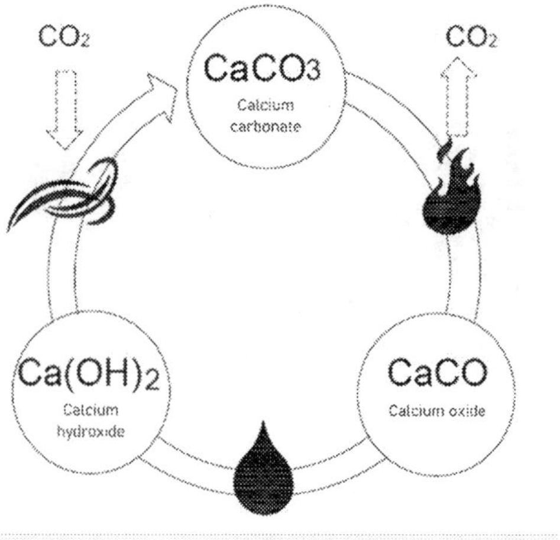

FIGURE 3.6 Lime cycle.
Source: Own elaboration.

TABLE 3.3 Lime Cycle Process

Elements	Materials	Chemical State	Chemical Formula	Weight (kg)
Earth	Limestone	Calcium carbonate	$CaCO_3$	100
Fire	• Boiling • Quicklime	• Releases water and carbon dioxide • Calcium oxide	CO_2 CaO	–45 = 55
Water	• Extinction • Slaked lime	• Extinction with water 33% • Calcium hydroxide	H_2O $Ca(OH)_2$	+18 = 73
Air	• Plaster: water, coloring • Carbon dioxide fixation • Water evacuation • Carbonated lime	• Water addition • Carbon dioxide capture • Water evaporation • Calcium carbonate	$+ H_2O$ $+ CO_2$ $- H_2O$ $CaCO_3$	+ 45 –18 100

Source: Anonymous, Traditional Arts (2020).

3.2.3 TYPES OF LIME

Limes classification: The calcium oxide, which is the quicklime, must be among its elements. There are different types of limes, depending on their use they can be classified into:

1. **Steel Lime:** It is a type of quicklime with at least 90% calcium oxide and very low content of silica, sulfur, and phosphorus.
2. **Chemical Lime:** It is a calcium hydroxide or hydrated lime of high concentration of Ca(OH)2, normally above 90%
3. **Food Grade Chemical Lime:** It is a particularity of the chemical lime, that, in addition to the high concentration of hydroxides, meets the standards of maximum content of heavy metal and harmful compounds for the food industry.
4. **Agricultural Lime:** Covers the entire range of limes, including their carbonate precursors, the particularity is that its application is like an agricultural land improver.
5. **Construction Lime:** It includes calcium hydroxide with or without magnesium, called hydrated limes with contents of 75% and 85% in said hydroxides and whose field of application is focused on the construction industry." (National Association of Lime Manufacturers, 2020).
6. **Aerial Lime:** Limes that are composed mainly of oxide or calcium hydroxide, slowly hardened by the absorption of the CO_2, they do not harden with water since their properties are not hydraulic (see Table 3.4; National Association of Manufacturers of Lime and Derivatives of Spain, 2015).
7. **Hydraulic Limes:** They are limes mixed with cement, blast furnace slag, fly ash or other suitable additions, these limes set with water, although CO_2 also influences this hardening purpose (see Tables 3.5 and 3.6) (National Association of Lime Manufacturers and Derivatives of Spain, 2015).

TABLE 3.4 Aerial Lime Designation in Function of the Content CaO + MgO

Aerial Lime Types	
Nomination	**Notation**
Calcium lime 90	CL 90
Calcium lime 80	CL 80
Calcium lime 70	CL 70
Dolomitic lime 85	DL 85
Dolomitic lime 80	DL 80

Source: National Association of Lime Manufacturers and Derivatives of Spain (2015).

TABLE 3.5 Hydraulic Natural Lime Type

Natural Hydraulic Lime Type	
Nomination	**Notation**
Natural hydraulic lime 2	NHL 2
Natural hydraulic lime 3,5	NHL 3,5
Natural hydraulic lime 5	NHL 5
Formulated Limes Types	
Designation	**Notation**
Formulated lime A 2	FL A 2
Formulated lime A 3,5	FL A 3,5
Formulated lime A 5	FL A 5
Formulated lime B 2	FL B 2
Formulated lime B 3,5	FL B 3,5
Formulated lime B 5	FL B 5
Formulated lime C 2	FL C 2
Formulated lime C 3,5	FL C 3,5
Formulated lime C 5	FL C 5
Hydraulic Limes Types	
Designation	Designation
Hydraulic lime 2	Hydraulic lime 2
Hydraulic lime 3,5	Hydraulic lime 3,5
Hydraulic lime 5	Hydraulic lime 5

Source: National Association of Lime Manufacturers and Derivatives of Spain (2015).

TABLE 3.6 Resistance to the Compression of the Hydraulics Limes and the Natural Hydraulic Limes

Types of Constructional Limes	Resistance to the Compression (MPa)	
	7 Days	28 Days
HL 2 and NHL 2	–	≥ 2 to ≤ 7
HL 3,5 and NHL 3,5	–	$\geq 3,5$ to ≤ 10
HL 5 and NHL 5	≥ 2	$\geq 3,5$ to ≤ 10

*If HL 5 and NHL 5 have a visible density underneath to 0,90 kg/dm^3, it is permitted that the resistance can reach up to 20 MPa

Source: National Association of Manufacturers of Lime and Derivatives of Spain (2015).

3.2.4 CHARACTERISTICS AND QUALITIES OF THE AERIAL LIME

The air lime is commonly used for the manufacture of paints with lime, its characteristics and qualities represent a direct impact in the final result of the paint to elaborate, therefore it is important to highlight these.

1. **Color:** The air lime with high content of calcium is white, and the purest the calcium is, the whiter the color will be. Unlike this, the hydraulic or dolomitic limes present a grayer coloration and therefore their use for paints is almost uncommon.
   Furthermore, the lime paint has a brightness and transparency that requires several layers of paint to achieve an even color on the surface. As an additive to improve the coverage of the product it is possible to add up grounds (Cerón and C., 2016).
2. **Curing:** "It is called curing to the chemical process in which the constitution water becomes supplanted by CO_2 from the atmosphere, in this way the calcium hydroxide returns to be calcium carbonate again" (Cerón and C., 2016):
   "Air curing is a slow process because it needs the presence of large amounts of CO_2 in the air for it to form the necessary carbonates and, by being produced from the outside to the inside, the first layers already carbonated make the passage of CO_2 to the more tender layers. This transformation needs a minimum level of relative humidity, the maximum speed of carbonation is obtained with 50%, if the humidity is too low amorphous calcium carbonate is produced" (Cerón and C., 2016).
3. **Permeability:** The air lime facilitates the transmission of humidity as one of its properties is that it has high breathability.
4. **Alkalinity:** The alkalinity that lime hydroxide possesses is higher than 12, this quality specifically offers disinfectant and sterilizing characteristics to the material, since it attacks the acidity of the living organisms. On the other hand, the high pH makes it impossible to use organic pigments, but not minerals. A disadvantage of this type of pigments is that its color varies slightly over time (Cerón and C., 2016).
   However, once the product has cured, the pH will be diminished to 8.3 approximately (Cerón and C., 2016).

5. **Benefits:**
   - Esthetic value in its white color, brightness, and transparency;
   - Improves the air quality because it absorbs CO_2;
   - Allows the transpiration of the surfaces on which it is placed, preventing moisture and molds that can cause damages to the users' health;
   - Contains disinfectant and sterilizing qualities because of its high pH.

3.2.5 LIME-BASED PAINT

3.2.5.1 DEFINITIONS OF PAINTS

Paint is defined as the mixture of solid pigments with a liquid excipient, applied as a thin layer over any surface (Ching and Binggeli, 2015).

All the paints are integrated by different based-compounds that provide their main physical and chemical characteristics, which are described in Table 3.7.

TABLE 3.7 Paint Components

Vehicle		**Pigments and Grounds/Fillers**		**Additives**
Binder	**Solvent**	**Pigments**	**Grounds/Fillers**	
Maintains all the components united, provides adherence, stability, and durability. Conditions the performance of the paint.	Provides the paint optimal consistency so that its use is easy to apply.	Give color and opacity to the paint. They can be natural or artificial, inorganic, or organic.	They are from inorganic nature; they are used to contribute solid matter to the paint. They also modify the paint's features, whether it is the viscosity, the rheology, and the brightness.	Its use is not mandatory; its function is to emphasize or provide other characteristics to the paint, it is applied in small proportions.

Source: Lévano Cerón (2016).

3.2.5.2 CLASSIFICATION OF THE PAINTS

There are various forms of classifying the paints, either from their components and qualities, or according to its function or use, among other things. In a very general way, they can be classified as follows:

1. **Synthetic Paints:** This type of paint is formulated with products derived from petroleum.
2. **Mineral Paints:** Developed with mineral binders, such as silicate and lime.
3. **Natural Paint:** This type of paint is formulated based on ingredients of plant or animal origin.

3.2.6 LIME-BASED PAINT CLASSIFICATION

In the wide variety of uses that lime has in the construction industry, lime-based paint will be used in this occasion, for being an aqueous solution that, as it was indicated in the beginning, it is one of the uses of the lime to combat virus in some diseases, and therefore it is important to see how lime-based paint can function to combat the infection by SARS-CoV-2. It will begin firstly by explaining how lime-based paint is made, as well as its main uses (see Table 3.8).

TABLE 3.8 Types of Paints and Their Applications

<table>
<tr><th>Type</th><th>Type of Lime</th><th>Use</th><th>Dose of Lime-Water (Volumes)</th><th>Coloring in Relation with the Lime</th><th>Consumption by Layer and m²</th></tr>
<tr><td>Liming</td><td>Aerial</td><td>Facade
Habitation
Decoration
Agricultural</td><td>1/2–1/3</td><td rowspan="2">Maximum 15% of synthetic oxide; 25% of natural land in relation to the weight of the lime.</td><td>300 gr</td></tr>
<tr><td>Patina</td><td>Aerial</td><td>Stones
Decoration</td><td>1/4–1/6</td><td>150 gr</td></tr>
<tr><td>Glaze</td><td>Aerial</td><td>Staining
Consolidation</td><td>1/6–1/10</td><td rowspan="2">Maximum 15% of synthetic oxide; 25% of natural land in relation to the weight of the lime, an additive can be incorporated.</td><td>100 gr</td></tr>
<tr><td>Lime water</td><td>Aerial</td><td>Consolidation</td><td>Water used in the shutdown of the decanted and filtered lime.</td><td>50 gr</td></tr>
</table>

Source: Lime and Plaster Craftsmen (2014).

The paint with lime can be classified in three groups, the so-called "lime-based paints," "lime paint" and "paint of lime"

They are conformed by:

- A lime binder, that can be aerial or hydraulic lime. This component conditions the performance of the paint.
- Natural land pigments, which must be applied in a maximum proportion of 25% of the weight of the cal. Or also it is possible to apply iron oxides, in a maximum proportion of 15% of the weight of the lime.
- A solvent, which can be water or lime water.
- Additives, which are applied only if it is necessary (Levano, Navarro, and Rosell, 2018).

There are two application techniques:

- By fresco or Buon Fresco, in this technique the pigments are diluted in lime water and subsequently integrated in a plaster when this one still is cool; and
- By drying, on this technique, the paint is applied on the dry substrate of mineral origin. In accordance with the proportion of binder and solvent composed in the formula, the type of paint can be classified as: glaze, paint, or liming/whitewashing (see Table 3.9).

TABLE 3.9 Lime and Solvent Volume for Lime-based Paints

Type of Lime-based Paint	Lime Volume	Water or Lime Water Volume
Glaze	1L	2–3 L
Paint	1L	4–7 L
Liming or whitewashing	1L	7–30 L

Source: Levano, Navarro, and Rosell (2018).

1. **Lime-based Paint:** Composed of:
   i. Lime;
   ii. Water or lime water;
   iii. Natural lands or iron oxides;
   iv. Organic additives, which should not exceed the stated content of heavy metals and volatile organic compounds in their dosing.
2. **Lime Paint:** Composed by:
   i. Lime;

ii. Water or lime water;
iii. Natural pigments such as iron oxides (see Figure 3.7);
iv. Non-organic additives (Levano, Navarro, and Rosell, 2018).

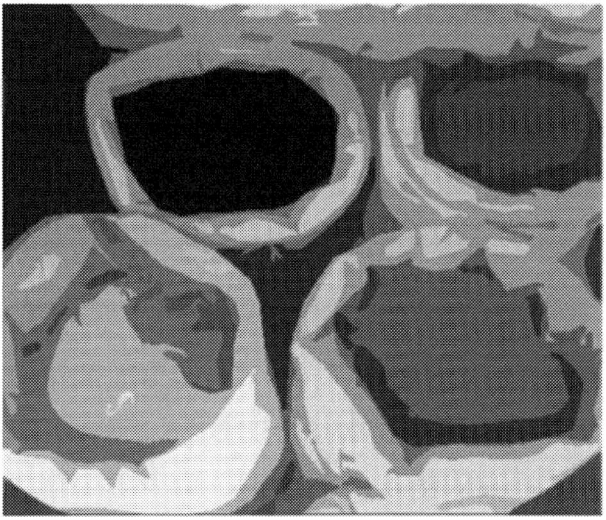

FIGURE 3.7 Iron oxide pigments.
Source: Natalia Franco drawing based on the original photograph; Levano, Navarro, and Rosell (2018).

The qualities found in handcrafted lime paint are given in Table 3.10.

TABLE 3.10 Handcrafted Lime Paint Qualities

Resistance Factor to Steam μ	**4–10 μ if a Waterproofing Additive Isn't Added**
Alkalinity	pH 12–13
Adherence	Cohesive mineral substrates including lime paint's old layers. By being carbonated, the paint is attached to the mineral surface and is petrified creating a micro-crystalline structure.
Biocide Action	Present due to its high alkalinity. This property lasts up until the end of the carbonation process.

Source: Levano, Navarro, and Rosell (2018).

Some particularities or characteristics regarding the finish and its consistency are the following:

- Pigmentation varies according to the force with which it is applied in each brush;
- When the paint dries, the color can be 50% lighter;
- The product tends to settle into its container, for which it must be constantly stirred during its application so that the components are properly integrated;
- It is necessary to apply 2 to 8 coats, depending on the density of the paint, to cover a surface evenly (Levano, Navarro, and Rosell, 2018).

3.2.6.1 DISADVANTAGES OF LIME PAINTS

The disadvantages or defects that may arise once the paint has been applied to the surface are generally due to poor execution. Besides, these types of paints also require greater care or maintenance than synthetic paints commonly used today.

Here are some of the most common defects that appear once the paint has been applied, their possible causes, and how these can be corrected or restored (see Table 3.11).

TABLE 3.11 Most Common Defects in Lime-based Paints

Defect	Cause	Restauration
The paint stains	Bad carbonation due to not pre-moistening the facing, very high temperatures and low RH, excess lime in the mix or saturation of the paint with pigment.	Application of consolidates. Example: lime water.
Loss of adherence	Poor substrate preparation (lack of pre-wetting), activation error or incompatible substrates.	Chopping the substrate and repainting, consolidation with lime water, adding glue, glues, or powdered resins or by injection.
Presence of clumps	Poor pigment preparation	Soak the pigment in water for three or four days and sift it to avoid clumps.
Fissures and cracks	Inadequate support or poor execution	Depending on the type of fissure, determine the defect and correct it.

Source: Cerón and C (2016).

3.2.6.2 HOW TO MAKE LIME PAINT?

Paints are substances that mix binders, solvents, and pigments used to protect and decorate interior and exterior surfaces. Conventional paints, unlike lime paints, are made with chemicals that are harmful to humans and the environment for which the importance of using natural paints where lime paints are found.

The best lime paints are those that are manufactured with quicklime (CL90), such as quicklime which has better quality than slaked lime or Calidra, the proportions of the material and other ingredients may vary according to the region in which it occurs. below is the description of the elements that make up these paintings.

3.2.6.3 PIGMENTS

If it is desired not to use the natural color of the lime paint, pigments must be incorporated to give it a color. These pigments must be of mineral origin that are compatible with the alkalinity of the lime, which will achieve greater tensile power. It must be taken into consideration that the colors of the pigments will have a degradation of up to 50% due to the white of the lime, it is therefore important that if you want to have more intense colors you should use additives, binders, and fixatives to achieve so (Lévano Cerón, 2016).

3.2.6.4 BINDERS

This element is the one that gives the paint adherence, stability, and durability to the lime; it is the substance that binds to the other components that the paint contains, so it is the one that mainly conditions its behavior.

3.2.6.5 SOLVENTS

The main function of this element is to make the paint more workable, with which it has a better penetrability and adherence in the pores of the surfaces, by improving its consistency.

3.2.6.6 ADDITIVES

This element, although it is not a product that must necessarily be incorporated into lime paints, if it is a substance that improves certain characteristics of the paint, it must be incorporated in small proportions into the mixture of the other elements mentioned above (see Tables 3.12 and 3.13) (Lévano Cerón, 2016).

TABLE 3.12 Current and Traditional Additives

	Type	Use	Inconvenience	Dosage
Current	**Binders:** Synthetic resins	Fix excess pigment and improve adhesion	Loss of porosity	0 to 2 l per 10 l of paint
	Water retainers: Methylcellulose	Slows dehydration and improves stability	Loss of porosity and cannot be used in cold weather	Maximum 50 gr per 10 l of paint
	Expansive: Soaps	It favors the dispersion of the pigment and the implementation of the painting	In excess, it causes foam and modifies the covering power. Better to avoid in light shades	Up to 10 gr per 10 l of paint
Traditional	**Binders:** Lactalbumin	Gouache paint fixer on plasters	Possible organic decomposition. Loss of porosity	100 gr. For 10 kg. lime paste
	Waterproofing: Oils	Wooden supports and baseboard waterproofing	Loss of porosity	10–20% of the weight of quicklime
	Accelerators: Alum	Accelerates carbonation	Loss of porosity and possible efflorescence	10% by weight of lime paste
	Expansive: Vinegar	Accelerates carbonation and improves commissioning	Alters the pH of lime	40 cm^3 per 10 kg of lime paste

Source: Enhancement of lime paint: Analysis and characterization of paint prototype (p. 11); Lévano Cerón (2016).

TABLE 3.13 Traditional Additives Complement

Additive	Use	Side Effect	Dose
Milk	Complementary binder, increased water resistance	Loss of porosity, color fixation	Maximum 15% by weight of lime
Casein	Complementary binder, higher water resistance, used on wood	Loss of porosity, fixation of colors	Maximum 5% by weight of lime
Egg-white	Albumin, increases resistance to outside and humidity	With some very caustic limes it is cooked	Maximum 5% by weight of lime
Egg yolk	Decoration, of great finesse and beauty, increased resistance to humidity	Alters whites, loss of porosity	Maximum 10% by weight of lime
Alginate	Resistance to moisture, elasticity. Moisture retainer	Slight loss of porosity, possible appearance of salts in humid places, greater transparency	Maximum 5% by weight of lime
Strong GLUE	Increases the ability to adhere and resistance to inclement weather, crystallizes allowing the passage of water vapor	Bad smell, yellowing. Greater transparency, slows down carbonation	Maximum 5% by weight of lime
Bone glue	Increases the ability to adhere and resistance to inclement weather, crystallizes allowing the passage of water vapor	Bad smell, yellowing. Greater transparency, slows down carbonation	Maximum 5% by weight of lime
Skin glue	Increases bonding capacity and resistance to inclement weather	Bad smell, yellowing. Increased transparency slows carbonation, creates film, and prevents the passage of water vapor	Maximum 5% by weight of lime
Rabbit-skin glue	Increases bonding capacity and resistance to inclement weather	Bad smell, yellowing. Greater transparency, slows down carbonation, creates film and prevents the passage of water vapor	Maximum 5% by weight of lime

TABLE 3.13 *(Continued)*

Additive	Use	Side Effect	Dose
Fish glue	Increases the ability to stick and resistance to bad weather, increases elasticity	Bad smell, yellowing. Greater transparency, slows down carbonation, creates film and prevents the passage of water vapor	Maximum 5% of the weight of lime
Flour	Reacts with lime to form a water-insoluble product, increases hardness. When mixed with olive or flax oil it gives very beautiful inks	Loss of porosity	Maximum 15% by weight of lime
Flour glue	Reacts with lime to form a water-insoluble product, increases hardness. When mixed with olive or flax oil it gives very beautiful inks	Loss of porosity. Increased transparency	Maximum 10% by weight of lime
Cornstarch glue	Reacts with lime to form a water-insoluble product, increases hardness. Mixed with olive or flax oil it gives very beautiful inks	Loss of porosity. Increased transparency	Maximum 10% by weight of lime
Rice glue	Reacts with lime to form a water-insoluble product, increases hardness. Mixed with olive or flax oil it gives very beautiful inks	Loss of porosity. Increased transparency	Maximum 10% by weight of lime
Starch-based glue	Reacts with lime to form a water-insoluble product, increases hardness. Mixed with olive or flax oil it gives very beautiful inks	Loss of porosity. Increased transparency	Maximum 10% by weight of lime

Source: Lime and Plaster Craftsmen (2014).

3.2.6.7 SUPPORTS OR SURFACES

Substrate or support is how the material or surface on which the paint will be applied is called. It is important to note that the chemical and physical components of the substrate must be compatible with the components of the coating to be applied. Therefore, lime-based paint cannot be applied to all substrates. Regarding the chemical characteristics of the support, these should not react negatively with the paint; and physically, the texture of the material must be porous for the product to adhere to it (Lévano Cerón, 2016).

Next, in Table 3.14, the types of substrates and their compatibility are presented according to their chemical and physical characteristics.

TABLE 3.14 Compatibility of Substrates with Lime-based Paint

Material	Chemical Compatibility with Lime Paint	Physical Compatibility with Lime Paint	Observations
Steel	Yes	No	It is a material with a very smooth surface, so its use as a substrate is not recommended.
Brick	Yes	Yes	It is compatible both chemically and physically. Its use as a substrate is recommended.
Concrete	Yes	Yes	It is compatible both chemically and physically. Its use as substrate is recommended.
Mortar	Yes	Yes	It is compatible both chemically and physically. Its use as substrate is recommended.
Rammed earth and other soil substrates	Yes	Yes	It is compatible both chemically and physically. Its use as substrate is recommended.
Zinc	No	–	It is not chemically compatible since the characteristics of lime react with those of zinc. Its use as substrate is not recommended.
Aluminum	No	–	It is not chemically compatible since the characteristics of lime react with those of aluminum. Its use as substrate is not recommended.
Synthetic Paintings	No	–	It is not chemically compatible since the characteristics of lime react with those of synthetic paintings. Its use as substrate is not recommended.
Plaster	Yes	No	Although we are talking about a porous material, its absorption capacity does not allow the paint to carbonize properly.

TABLE 3.14 *(Continued)*

Material	Chemical Compatibility with Lime Paint	Physical Compatibility with Lime Paint	Observations
			However, it is used as a substrate as long as an additive is added to the paint that serves as a moisture retainer or another that can generate an adhesion between the paint and the support.
Metal	Yes	No	It is not physically compatible due to its nonporous surface, adhesion can be improved with compatible primers, however, durability will be affected.
Plastics	Yes	No	It is not physically compatible due to its nonporous surface, adhesion can be improved with compatible primers, however, durability will be affected.
Glass	Yes	No	It is not physically compatible due to its nonporous surface, adhesion can be improved with compatible primers, however, durability will be affected.

Source: Lévano Cerón (2016).

As seen previously, it is recommended that the surfaces or supports for the application of lime paints be porous, so that the lime can penetrate into the pores of the mentioned surfaces, in some occasions this is not possible so additives must be used to make the bond between the painting and the surface or support, chart 15 below is a description of the different types of support and what their drawbacks are (Table 3.15).

TABLE 3.15 Types of Supports and Their Characteristics to Receive Lime Paintings

Surface or Support	Directly	Attached	Additive	Drawback
Porous calcareous stone	Yes	No	No	Is an appropriate material for liming
Nonporous stone	No	Yes	Yes	Use hydrochloric acid, use an additional complimentary binder (glues or resins) moisture retainers used in the mix or on the support

TABLE 3.15 *(Continued)*

Surface or Support	Directly	Attached	Additive	Drawback
Brick	Yes	No	No	Appropriate material
Mud	Yes	Yes	Yes	Very porous, use moisture retainers, supplementary binders, and plasticizers.
Old plaster	Yes	Yes	Yes	The old plasters have less porosity and were more suitable to use additional complementary binder (glues or resins) to retain moistures
Modern plaster	No	Yes	Yes	It is hard to whitewash, to use binders and additives
Mortars	Yes	Yes	Yes	If they have not been waterproofed, after cleaning they are suitable
Woods	Yes	Yes	Yes	Use binders and adhesives, casein, etc., applied in the mass or on the wood. Sand, wet, and paint
Metals	No	Yes	Yes	Different coefficient of expansion, paint fall over time, use adhesives and mass additives
Cements	Yes	Yes	Yes	Clean and remove salts
Drywall	No	Yes	Yes	Very porous, use moisture retainers, complementary binders, and plasticizers.
Plastics	No	Yes	Yes	Different coefficient of expansion, paint fall over time, use adhesives and mass additives.
Glass/Crystal	No	Yes	Yes	Etch the glass with acid use mass additives or on the contact surface
Ancient paintings	Yes	Yes	Yes	Use additives to apply on liming
Modern paintings	No	Yes	Yes	Use adhesives, make research to give the right solution

Source: Lime and Plaster Craftsmen (2014).

3.2.7 TYPES OF LIME PAINT

The variant of the types of lime paint is found in the type of lime used for its manufacture, so according to this parameter we can find the following types of paints:

1. **With Aerial Limes:** Calcium oxide CL90 is used in powder or paste, which must have been in water for at least 6 months, the lime must be of high purity, and it can be used until the third day of its preparation.
2. **Hydraulic Limes:** These have the particularity that they harden faster than the aerial ones, therefore there may be difficulties in their manufacturing process for this reason, the main characteristic of these limes used to make paintings is that they have a greater setting of colors.
3. **Synthetic Limes:** Using this type of lime for the production of paintings, the recommendations for paintings made with hydraulic limes must be followed.

3.2.8 MANUFACTURE OF LIME PAINT

As already mentioned in the point 1.4.5.1, to be able to make lime paint; pigments, a binder, a solvent, and a fixer are required. Below in chart 8, a comparison is made about the proportions of each element for the manufacture of lime paint made by different institutions in Mexico (see Tables 3.16 and 3.17).

TABLE 3.16 Comparison of Methodologies for the Implementation of Paint based on Cactus Resin

Source	Ingredients	Observations
PROFECO	5 large nopal pads (30×20 cm)	Low quality and low hiding power.
	2.5 kg of lime	
	2 cups of kitchen salt	
	Color cement stain of your choice (the amount depends on the intensity of color you want to obtain)	
	6 L of water	

Eco-techniques – Secretary of the Environment	5 to 7 kg of hydrated lime. 2 to 2.5 kg of white cement. 5 to 7 large and preferably ripe nopal stalks 0.5 kg of salt If any color is desired, it should be purchased from building materials stores.	Dense paint, high hiding power to the first hand
Cultural Institute of León	1 kg of quicklime 4 l of water 4 large and fleshy nopales 1 cup of kitchen salt Color: Paint with white nopal resin due to lime. If you want other shades, you can add cement stain	Very diluted paint, good white color tone, low hiding power.
Painting my house (painting and decoration)	4 to 5 large nopales 4 kg of slaked lime (building lime) 4 L of water, 1 cup of salt or water sealant fixer	Low-quality paint, it comes off easily.
Digital library CONEVyT	1 kg of lime 4 l of water 2 l of nopal resin 1 cup of kitchen salt Cement colorant (depending on the intensity of the color, it is the amount of colorant)	High covering power secondhand. It does not come off easily.

Source: Suggestion of a semi-automatic prototype for the elaboration of an ecological paint based on nopal (p. 17); Aguilar Valencia, Hernández González and López Orihuela (2016).

TABLE 3.17 Other Methodologies for the Implementation of Paint based on Nopal Resin and Without Nopal

Source	Ingredients
Hydrated lime	9 kg of lime 19 l of water 0.250 Kg of kitchen salt 1 l of vinyl sealer 0.1 to 0.15 Kg of mineral pigment
Prof. Xóchitl Mora Gómez Dr. Lorena Vargas Rodríguez	10 Kg of lime 8 l of water 0.035 Kg of kitchen salt 0.800 Kg of nopales

Sources: Practice guide for construction (p. 104); Tovar Alcázar (2019); and white walls, a worthy alternative: natural paint based on lime and nopal (p. 14); Mora Gómez and Vargas Rodríguez (2015).

Next, the process for the elaboration of the two paints that gave the best result in the comparison is described in Table 3.8.

1. **Eco-Techniques Secretary of the Environment:**
   i. **Ingredients:**
      - 5 to 7 kg of calhidra;
      - 2 to 2.5 kg of white cement;
      - 5 to 7 large and preferably ripe nopal stalks;
      - 0.5 kg of salt.

   ii. **Preparation:** In an 18-liter bucket of water, the 5 or 7 stalks of previously chopped nopal should be placed, and it should be left for three days to release the mucilage, having to cover it in the indicated time.

   Later, the 5 to 7 kg of hydrated lime and the 2 to 2.5 kg of white cement and the 0.50 kg of table salt will be added, during the incorporation of the ingredients, it must be stirred constantly, so that groups are not created; if you want the paint to have color, cement stain will be added, as it does not alter the properties of the paint.

   If you want to prepare a sufficient amount of paint and that the shade does not change, you should consider to prepare the paint in a large container such as a 200-l drum, considering that the amounts of the ingredients should be adjusted so that proportions are conserved (Aguilar Valencia, Hernández González, and López Orihuela, 2016).

   The application will be made with a brush or a roller.

2. **Digital Library National Council of Education for Life and Work (CONEVyT):**
   i. **Ingredients:**
      - 1 kg of lime;
      - 4 l of water;
      - 2 l of nopal resin;
      - 1 cup of kitchen salt.

   ii. **Utensils:**
      - A container with a capacity of 5 L;
      - Plastic container with a capacity of 19 L;
      - Plastic or wooden shovel;
      - Strainer.

   iii. **Preparation:** First, three to four nopales are chopped (they are not vegetable nopales, but forage nopales, it can be Opuntia

Ficus-indica) and they are placed in a container with two liters of water and left to rest overnight in order to release the mucilage, once this is done, a strainer is taken and only the mucilage thus obtained is recovered.

In a 19-liter bucket, the lime must be mixed with water, and the salt is added, mixing perfectly well to avoid the formations of lumps, then the nopal mucilage is added, stirring it constantly so that it is perfectly integrated into the mixture of lime, water, and salt; in this step you can add colorant so that the paint is not white and let it rest for one more night and then, the paint is ready to use.

3. **Paint Without Nopal:**
   i. **Ingredients:**
      - 9 kg of lime;
      - 19 l of water;
      - 0.250 kg of kitchen salt;
      - 1 l of vinyl sealer;
      - 0.1 to 0.15 g of mineral pigment.

   ii. **Preparation:** In a 19 L tray with water 9 kilos of lime are added. Its incorporation has to be done by constantly shaking it so that lumps do not get formed, to subsequently let it rest for 2 hours. Once that time has passed, the salt and the liter of vinyl sealer are incorporated. This last one is not indispensable.

3.2.9 GRAPHIC DESCRIPTION OF THE PRODUCTION PROCESS OF LIME PAINT (FIGURE 3.8)

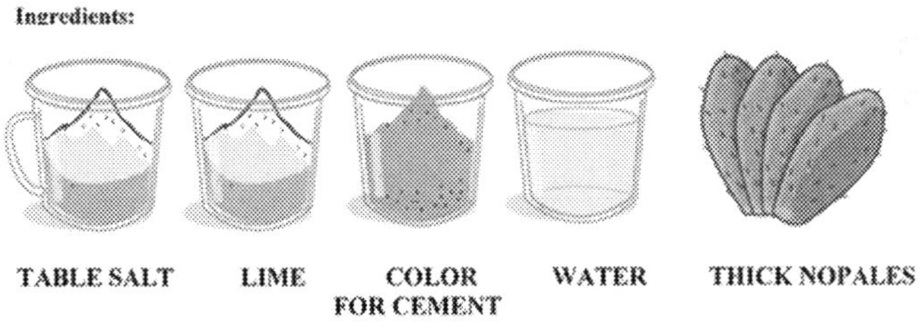

FIGURE 3.8 *(Continued)*

Nopales' preparation (Opuntia ficus-indica):

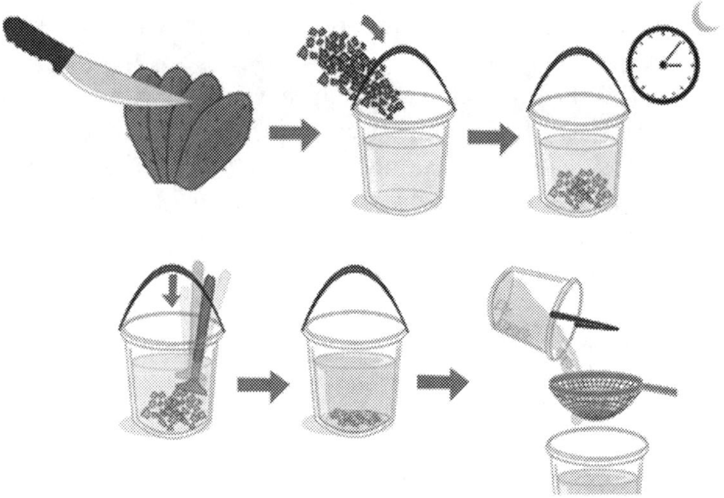

Chop the nopales in small pieces and place them in a tray with water, let it rest for 8 to 12 hours, afterwards mash the pieces to extract the mucilage and finally strain the water with mukage

Mixing of lime water and salt:

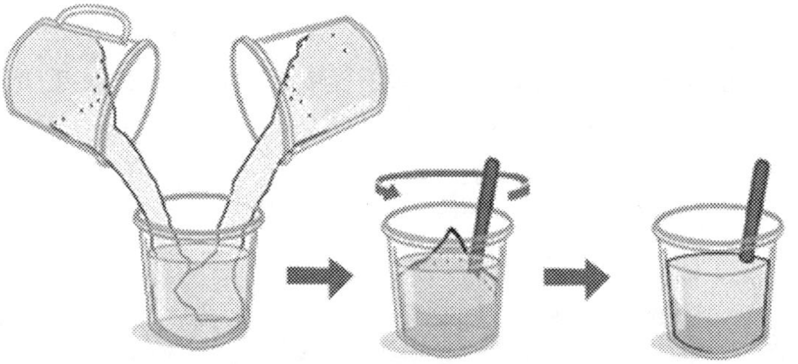

In a tray put 2 liters of water and incorporate the lime and the salt in the water, constantly shaking it so that lumps don't get formed until a homogeneous mix remains.

FIGURE 3.8 *(Continued)*

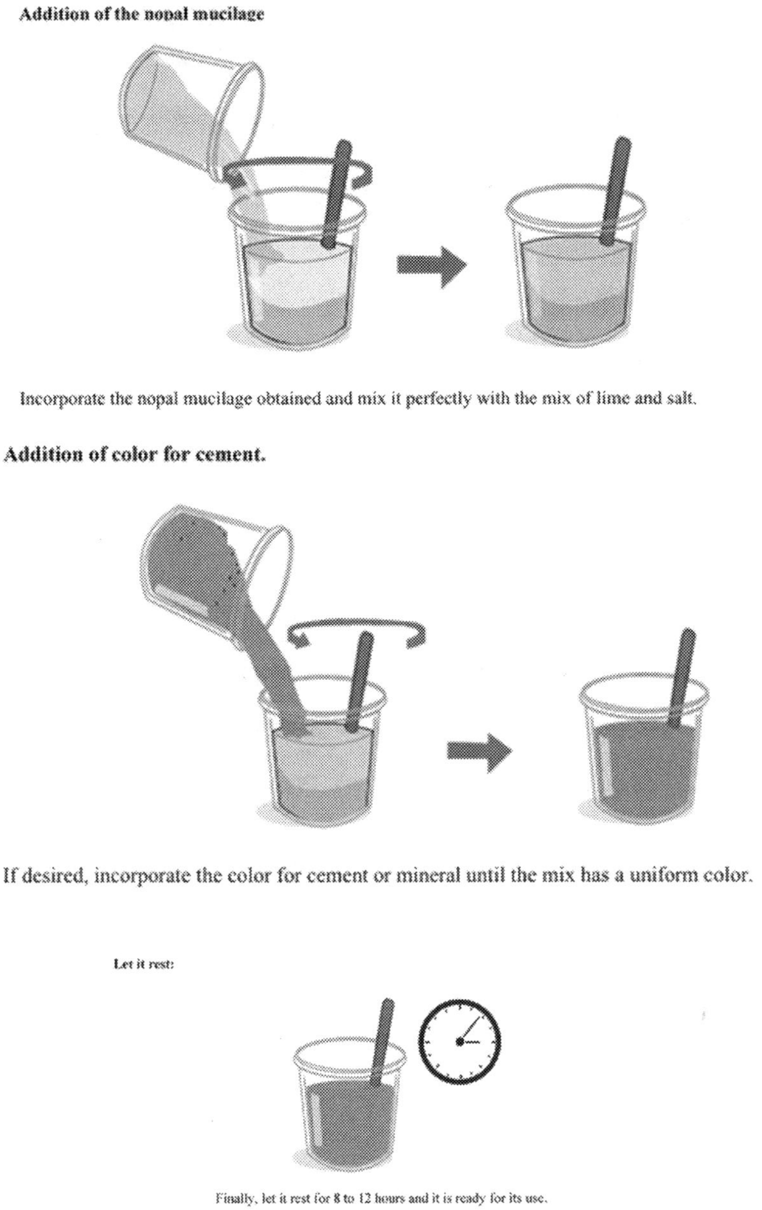
Addition of the nopal mucilage
Incorporate the nopal mucilage obtained and mix it perfectly with the mix of lime and salt.
Addition of color for cement.
If desired, incorporate the color for cement or mineral until the mix has a uniform color.
Let it rest:
Finally, let it rest for 8 to 12 hours and it is ready for its use.

FIGURE 3.8 Graphic description of the elaboration process of lime paint with nopal mucilage.
Source: Its own elaboration is based on the preparation of ecological nopal paint with lime; Pigmacolor (2020).

3.2.10 LIME PAINT'S TECHNICAL DATA

Next, the technical information of the lime paint obtained from investigations of this type of paint is presented (see chart 13), where data about the setting of pH, viscosity, density, the setting of the non-volatile material and the resistance to steam are presented. Next, the procedures for the tries of each one of the tests fulfilled to the lime paint are presented.

3.2.10.1 PH DETERMINATION

To test the pH, test strips will be used, which are immersed in the paint and when they change the coloring, it is proceeded to compare them with a scale of colors and by that determine the pH the paint has (see Figure 3.9) (Lévano Cerón, 2016).

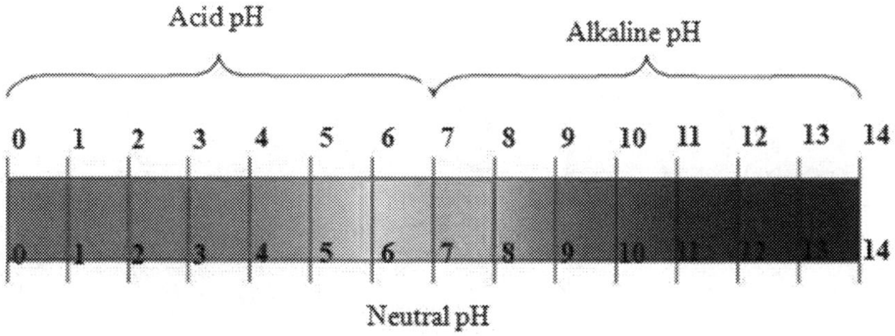

FIGURE 3.9 pH scale.
Source: Own elaboration, based on the information of Luque (2020).

3.2.10.2 VISCOSITY

To determine the viscosity, the next procedure was used: A viscometer is used to whip the paint to the following rpm 3, 4, 5, 6, 10, 12, 20, 50, 60, 100, 200 min-1. The whipping time was of 120 seconds, having to take care that the test temperature is of 20 ± 5°C (68–41°F), for the results' reading a software such as the Software Thermo Scientific HAAKE RheoWin can be used (Lévano Cerón, 2016).

3.2.10.3 DENSITY

To make this test, a pycnometer needs to be used (see Figure 3.10), which will be measured with distilled water. The empty recipient needs to be weighted and, when it is full of water, it has to be at a temperature between 22 to 23°C (71.6 to 73.4°F), the recipient's volume needs to be calculated.

To determine the paint's density the stipulated procedure noted in the last paragraph should be followed, replacing the water with the paint and the following formula will be applied:

$$\rho = \frac{m_2 = m_1}{V_t}$$

In which; ρ is the density; m_1 is the weight of the empty pycnometer; m_2 is the weight of the pycnometer with paint; V_t is the volume of the pycnometer.

For the test, the water and paint temperature should be between 22 to 23°C (71.6 to 73.4°F) and it should use pycnometers of 20, 50, 100, and 200 ml (see Figure 3.7) (Lévano Cerón, 2016).

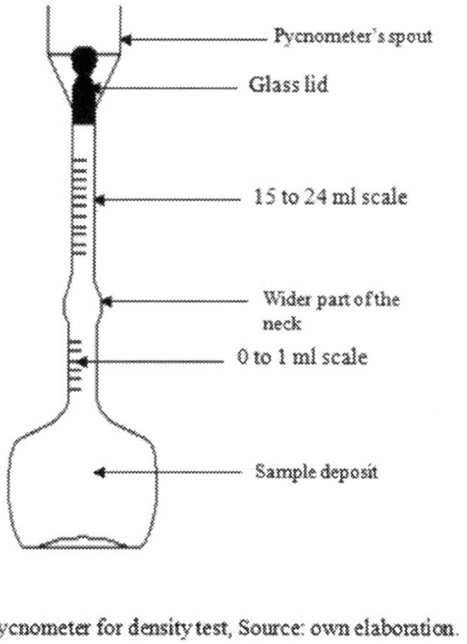

FIGURE 3.10 Pycnometer for density test.
Source: Own elaboration.

3.2.10.4 NON-VOLATILE MATERIAL DETERMINATION

For this test a glass recipient of 20 $cm^2 \pm 2$ cm^2 area will be used, which is weighted in the air and water. Next, the paint is applied by immersion and the painted recipient is weighted again, it is left to dry for 10 to 15 minutes at room temperature and once the above mentioned is done, it is put to dry in the oven for 10 minutes to a temperature between 70 and 80°C (158 and 176°F) to cure it a temperature of 125°C ± 5°C is maintained (257 ± 41°F) to finally let it cool at a room temperature in a desiccator, for it to be weighted in the air and water again.

The formulas used to determine the non-volatile material are the following:

a. For the non-volatile material in mass:

$$Nv_m = \frac{m_4 - m_1}{m_3 - m_1} \times 100$$

b. For the non-volatile material in volume:

$$Nv_v = Nv_m \frac{P_2}{P_p}$$

In which; m_1 is the mass of the recipient without paint weighted in the air in grams; m_3 is the mass of the recipient painted and wet, in grams; m_4 is the mass of the recipient painted and dried in the air, in grams; P_2 is the paint density in liquid state to the test temperature, in gr/cm^3; P_2 is the paint density in liquid state to the test temperature, in gr/cm^3; Nv_m is the medium content of the not volatile matter in percentage of the mass.

3.2.10.5 THE STEAM RESISTANCE FACTOR

To make this test, the paint samples are exposed to a moist environment in one of the sides of the sample and the other one will be in a dry environment. The difference in pressure makes the steam spread in the covering that it is being tested, this will cause a change of the mass in all the samples, which will be registered daily for a week.

A minimum of three test tubes should be tested, which are comprised by a slightly absorbent net into which three coats of lime paint will be applied, the environmental conditions to realize this test should be: An environmental temperature of 23°C (73.4°F) and a relative humidity of 473%, the samples must be left to rest for a week before doing the test.

Lastly, in Table 3.18, the criteria that the lime must accomplish according to the European regulation are presented.

TABLE 3.18 Technical Information about Lime Paint

Characteristic	Datum	Unit
pH determination	12–13	%
Viscosity	0.044	Pa.s
Density	1.25–1.67	gr/cm^3
Non-volatile material determination.	50–60	%
Steam resistance factor (μ)	12–60	–

Source: Lévano Cerón (2016).

3.3 LIME PAINT AS A SANITIZER

As it was mentioned in Sections 1.4.1 and 1.4.2, the WHO (World Health Organization) has determined that the substances that function as sanitizing should have a pH equal or superior to 12. On the other hand, the French laboratory Sanofi Pasteur has made tests and has verified that the virus of COVID-19 is eliminated within 24 hours in substances with a pH between 13–13.5% (Calcinor, 2020) moreover an important fact is that the lime paints elaborated with nopal mucilage increase their pH, since the mucilage has a pH of 5% (Río López, 2016), which makes the paints to be above the suggested pH. Furthermore, it is important to state that the effect of the approved substances by the WHO and presented in chart 1 of this document, indicate an effective durability of half a minute and maximum 10 minutes, which is different in comparison with the lime paints since the loss of pH is decreasing as the paint's carbonation effect is achieved, which is slow because of the amount of CO_2 that this absorbs. In studies already completed, it is observed that the minimum time in which these types of products reach, between a 60 and 70% of carbonation, is of 12 months (Luque et al., 2006) indicating that in this time the paint will have a pH between 8.1 and 9.5% on the basis that the original pH was of 13.5%, which indicates the time's effectiveness of lime paints, over the substances mentioned in Table 3.1, is greater and in consequence the effect as a sanitizer is more enduring.

Also, as it was mentioned before, in a region where the economies are not good and by facing the effects of the current pandemic, they have been more affected, being able to use a product such as lime paints to sanitize homes at a low cost is important.

As it is mentioned, the lime has been used in various processes as an element of sanitization for varied diseases.

As it was already mentioned, the lime in water dissolution increases its pH up to 13.5% and this makes it a substance with disinfection qualities by having antimicrobial, antiparasitic, biocidal, and mineralizer properties. This is achieved by the release of hydroxyl ions, which are radically reactive and fatal to most types of bacterias, fungus, virus, and nematodes (Calidra, 2020).

The lime's use is classified among the chemical disinfectants, generally this is applied in liquid state, in the case of lime there are lime paints that can be used for their big disinfectant impact, since it has been proved that it helps to block certain functions of the microorganisms that lead to its death. There is evidence that in some places it has been used to sanitize tools and equipment that has been contaminated with some harmful substances.

Lime is considered an ecological product since it is innocuous as much as for plants, food products, animals, and the same human being due to its low toxicity. In addition, it is a very cheap product and easy to obtain in any part of America.

The National Directorate of Animal Health of Mexico has published that lime is a very effective product for the eradication of diseases such as:

- Foot-and-mouth disease;
- Classical swine fever;
- Bovine viral diarrhea;
- Equine encephalitis;
- Transmissible gastroenteritis;
- Influenza A, B, C;
- Newcastle;
- Respiratory syncytial virus;
- Rinderpest;
- Enzootic bovine leukosis;
- Equine infectious anemia;
- Infectious bronchitis;
- Aujeszky's disease;
- Coital exanthema;

- IBR-IPV;
- Gallid alpha herpesvirus 1;
- Marek's disease;
- Smallpox;
- African swine fever virus.

Lime has also been used as an antiemetic since it controls cytotoxic vomit in children, it can be concluded that it is a product that has shown throughout its existence its efficiency to control bacteria and viruses in different areas of human life in addition to being a material that since its origin has been used like a material of construction but in the face of the current pandemic results in an alternative viable of use to disinfect or sanitize the interiors of the houses and thus guarantee the cleanliness of people's space.

3.4 CONCLUSIONS

In the present document, Calcium oxide or quicklime has been seen, since its origins as a material for construction and humanity along history has made use of this material for different activities in its life, as it was mentioned like a construction material, but also as an element to preserve food and water or to carry out food processes that allows us to obtain by-products of vegetable food like it is the case of corn and as by-product the dough, but also it is important to highlight its importance like a substance to control viruses and bacteria that cause serious diseases.

Although perhaps rarely mentioned throughout the document it is important to highlight that it is a product with sustainable features. This can clearly be seen in the analysis of the lime cycle since finally, lime tends to turn back to limestone when it absorbs CO_2 which makes this material to be unique for this characteristic that allows it to regulate CO_2 that exists in the atmosphere when integrated into its carbonation process.

So, for one part, its impact is double, it can regulate the proliferation of bacteria and viruses in the places where it is applied, and on the other the fact of being able to consume the CO_2 for a long time to transform into rock like it was already mentioned. In relation to the current situation in which it was also already mentioned, where it is important that the most disadvantaged population of Latin-American countries can have an inexpensive and effective material since those that have been suffering the most from the pandemic are the sectors of the population with low resources, it is known that they do not have medical services and generally live in unsanitary conditions, which increase the risk of contagion in this pandemic.

Thus, this job tries to offer this health alternative for these under-resourced communities and in itself for the whole community to know that there are viable and inexpensive alternatives to sanitize and at the same time have less impact on the environment and to use this product with these characteristics that have been mentioned throughout the document, in the hopes that it can be of use to our populations.

ACKNOWLEDGMENTS

I appreciate the participation of the student Natalia Franco Ruiz of the architecture degree and the students and teachers of the Language Office from Tamaulipas Institute of Higher Studies, A. C., who helped with the realization of this document.

KEYWORDS

- **carbon dioxide**
- **COVID-19**
- **gross domestic product**
- **literary review**
- **public health emergency of international concern**
- **World Health Organization**

REFERENCES

Aguilar, H., (2020). *History of Lime and its Properties Against Viruses*. Obtained from: Honorario Aguilar Architecture Studio: https://www.honorioaguilar.es/historia-de-la-cal-y-sus-propiedades-contra-los-virus/(accessed on 21 December 2021).

Aguilar, V. J. C., Hernández, G. C. A., & López, O. J. A., (2016). *Proposal of a Semi-automatic Prototype for the Elaboration of an Ecological Paint Based on Nopal*. México: IPN.

Anonymous, (2020). *2020 Coronavirus Disease Pandemic in Mexico*. Obtained from: https://es.wikipedia.org/w/index.php?title=Pandemia_de_enfermedad_por_coronavirus_de_2020_en_M%C3%A9xico&oldid=128860432 (accessed on 21 December 2021).

Anonymous, (2020). *Traditional Arts*. Obtained from: http://www.estucos.es/?page_id=1525 (accessed on 21 December 2021).

Barba, P. L., & Villaseñor, A. I., (2013). *Lime, History, Properties, and Use.* Mexico: UNAM.

Calcinor, (2020). Calcinor.com. Obtained from: https://www.calcinor.com/es/actualidad/corporativas/duplicado-usos-de-la-cal-en-la-lucha-contra-las-pandemias-13041 (accessed on 21 December 2021).

Calidra, (2020). *Calidra Always There.* Obtained from: https://calidra.com/todo-lo-que-necesitabas-saber-sobre-la-desinfeccion-con-cal/?gclid=EAIaIQobChMI9o6m3LT26wIVDfDACh2X_wiuEAAYASAAEgIB0fD_BwE (accessed on 21 December 2021).

Ching, F. D., & Binggeli, C., (2015). *Interior Design: A Manual.* Barcelona: Gustavo Gili.

Cymper, (2015). *Types of lime used in construction.* Obtained from: Cymper: https://www.cymper.com/blog/tipos-de-cal-utilizadas-en-la-construccion/(accessed on 21 December 2021).

European Lime Association, (2009). *Practical Guide on the use of Lime in the Prevention and Control of Avian Flu, Foot-and-Mouth Disease and other Infectious Diseases.* Brussels: European Lime Association.

García, G. J. C., (2019). Nixtamalization of corn: physicochemistry in action. *Reviata Ondícula*, 15–17.

General Directorate of Mining Development, (2018). *Chalize Market Profile.* Mexico: Secretary of Economy.

Google, (2020). *Google News.* Obtained from: https://news.google.com/covid19/map?hl=es-419&gl=MX&ceid=MX%3Aes-419 (accessed on 21 December 2021).

Guerrero, H. C. J., (2001). Limestone rocks: Formation, carbon cycle, properties, applications, distribution, and perspectives in the Mixteca Oaxaqueña. *Temas de Ciencia y Tecnología*, 3–14 (Science and Technology Topics, 3–14).

HealthyChildren.org. (2020). *New Coronavirus 2019.* Obtained from: https://www.healthychildren.org/spanish/health-issues/conditions/covid (accessed on 21 December 2021).

Informant, E., (2020). *The informant Mx.* Obtained from: https://www.informador.mx/Sufre-economia-de-Mexico-una-caida-historica-en-segundo-trimestre-del-ano-l202008270001.html (accessed on 21 December 2021).

KAHRS, R., (1995). General principles of disinfection. *Revue Scientifique et Technique-Office International des Epizooties*, 143–163.

Levano, B., Navarro, A., & Rosell, J., (2018). *Lime-Based Paints: Market Review and Regulatory Limits.* SciELO Analytics.

Lévano, C. B. (2016). *Enhancement of Lime Paint: Analysis and Characterization of Paint Prototype.* Barcelona: Escola Politècnica Superior d'Edificació de Barcelona – Universidad de Catalunya (Higher Polytechnic School of Edification of Barcelona-University of Catalunya).

Lévano, C. B. C., (2016). *Enhancement of Lime Paint: Analysis and Characterization of Paint Prototype.* Barcelona: UPC.

Lhoist, (2018). *Lime Throughout History: Lhoist.* Obtained from: Lhoist Web Site: https://www.lhoist.com/es/la-cal-lo-largo-de-la-historia (accessed on 21 December 2021).

Lime and Plaster Craftsmen, (2014). WordPress.com. Obtained from: https://artesdelascalesylosyesos.files.wordpress.com/2014/11/encalado-en-doc.pdf (accessed on 21 December 2021).

Luque, A., Sebastian, E., De La Torre, M. J., Cultrones, G., Ruíz, E., & Urosevic, M., (2006). Comparative study of lime mortars in paste and lime in powder. Carbonation control. *XXVI Reunión (SEM)/XX Reunión (SEA)-2006*, (pp. 293–296). Jaen: US.

Luque, F., (2020). *Scientific Experiments*. Obtained from: https://www.experimentoscientificos.es/ph/escala-del-ph/(accessed on 21 December 2021).

Mellaart, J., (1965). *Earliest Civilizations of the Near East Library of Early Civilizations McGraw-Hill Paperbacks*. McGraw-Hill.

Mora, G. X., & Vargas, R. L., (2015). Hite walls, a worthy alternative: natural paint base lime and nopal. In: *20th National Meeting on Regional Development in Mexico* (p. 21). Cuernavaca: UNAM.

Muñoz, R. C., Collazo, P. A., & Alvarado, F. J., (1995). Bactericidal effect of hydrated lime in aqueous solutions. *Bulletin of the Pan American Sanitary Bureau*, 302–306.

National Association of Lime Manufacturers, (2020). ANFACAL. Obtained from: http://anfacal.org/datos-tecnicos-de-la-cal-y-sus-derivados/(accessed on 21 December 2021).

National Association of Manufacturers of Lime and Derivatives of Spain, (2015). CYMPER.COM. Obtained from: https://www.cymper.com/blog/tipos-de-cal-utilizadas-en-la-construccion/(accessed on 21 December 2021).

National Network of Teachers of Traditional Construction, (2018). *Lime Teachers*. London: INTBAU.

OMS, (2020). *Questions and Answers on Coronavirus Disease COVID-19*. Obtained from: https://www.who.int/es/emergencies/diseases/novel-coronavirus-2019/advice-for-public/q-a-coronaviruses?gclid=EAIaIQobChMIyOK2xtjB6wIVS73ACh3nqgGcEAAYASAAEgJQGvD_BwE (accessed on 21 December 2021).

Palma, V., (2009). History of lime production in the North of Mexico Basin. Obtained from: CIENCIA ergo-sum, *Multidisciplinary Scientific Journal of Prospective* https://www.redalyc.org/pdf/104/10412057002.pdf (accessed on 21 December 2021).

Pan American Health Organization/World Health Organization, (2020). *Response from the PAHO/WHO. Report of August 3, 2020*. Washington, D.C.: WHO.

Pigmacolor, (2020). *Pigmacolor.com.mx*. Obtained from: http://www.pigmacolor.com.mx/assets/preparacion-ilustrada-pintura-de-nopal.pdf (accessed on 21 December 2021).

Practical solutions, (1994). *Technical Sheets*. Obtained from: https://www.emilio.com.mx/pdf/practicalaction/cal.pdf (accessed on 21 December 2021).

Río, L. E., (2016). *Technical-Economic Characterization of the Application of Lime-Based Paint for low*. Income housing in the State of Aguascalientes. Aguascalientes: UAA.

Ruiz, C. M., Ponce, A. C., & Alvarado, F. J., (1995). Bactericidal effect of hydrated lime in aqueous solution. *Pan American Journal of Public Health*.

Salud, S. D., (2020). *Coronavirus (COVID-19-Daily Technical Releases-August 2020)*. Obtained from: https://www.gob.mx/cms/uploads/attachment/file/574971/Comunicado_Tecnico_Diario_COVID-19_2020.08.28.pdf (accessed on 21 December 2021).

Shuña, M. G., & Lozano, J. T., (2016). *Antibacterial Capacity of Four Disinfectants*. Iquitos: UNAP.

The United States Environmental Protection Agency, (2020). *Use of Desinfectants and Coronavirus (COVID-19)*. Obtained from: https://espanol.epa.gov/sites/production-es/files/2020-04/documents/2020-03-26_-_lista_n_productos_con_declaraciones_de_patogenos_virales_emergentes_y_coronavirus_humano_para_usar_contra_sars-cov-2_fecha_27pp.pdf (accessed on 21 December 2021).

Tovar, A. R., (2019). *Practical guide for construction*. Mexico: Calidra.

Usedo, V. R. M., (2015). *Study and Analysis of the Use of Lime for Architectural Heritage*. Valencia: Polytechnic University of Valencia.

World Health Organization, (2004). *WHO Model Form 2004*. Spain: Ars Medica.

CHAPTER 4

Affordable Housing Resilient Design in Healthy Environments

ROLANDO ARTURO CUBILLOS GONZÁLEZ

Faculty of Design of the Catholic University of Colombia, El Claustro Headquarters – Bogotá, Diagonal 46 A # 15 B – 10, Colombia, E-mail: racubillos@ucatolica.edu.co

4.1 INTRODUCTION

COVID-19 pandemic is the most important event of the last year. This phenomenon demonstrated several limitations and weaknesses of our societies. But, in turn, opportunities are seen to improve these kinds of barriers in society. For example, one of these opportunities is the study of improvements in the quality of the environments of low-income people living in affordable housing. There is also an opportunity for improvement in the aspects that ensure that housing is healthy and safe against phenomena such as those of a pandemic (see Figure 4.1).

For example, empirical evidence is identified because of the COVID-19 pandemic, and it is necessary to reflect on how the spaces in which we live are designed (Tokazhanov et al., 2020). These changes are closely linked to design processes, standards, and of course, sustainability requirements. This is especially true in the case of housing and particularly accessible housing. That because of its characteristics of belonging to a low-income population is more vulnerable to such pandemic phenomena. In principle, three key factors have been identified when making changes to accessible housing, these are (see Figure 4.2): (i) flexibility; (ii) building quality; and (iii) quality of life (Cubillos-González, 2014, 2017).

Architecture for Health and Well-Being: A Sustainable Approach. Maria Eugenia Molar Orozco, PhD (Ed.)

FIGURE 4.1 Housing is healthy and safe against the pandemic, 2017.
Source: Photographs of the author.

FIGURE 4.2 Key factors have been identified when making changes to accessible housing, 2017.
Source: Photographs of the author.

In this sense, the need for the design of homes that offer health protection to their users is identified, such as the introduction of automated technologies that reduce and control the contact of agents outside the house. In addition to the introduction of external infection risk control systems. As well as the need to ensure that designed spaces reduce the risk of increased mental health problems due to isolation for extended periods. The next chapter aims to build a theoretical framework of reference on the relationship between affordable housing, resilient design, and healthy environments within the context of the COVID-19 pandemic (see Figure 4.3). To this end, a literary review was carried out to identify the most relevant factors concerning these issues and research. Then, the constructs that explain each concept are presented. Finally, the conclusions are presented to strengthen the study of the design of affordable homes in the context of the COVID-19 pandemic.

FIGURE 4.3 Accessible housing, resilient design, and healthy environments, 2017.
Source: Photographs of the author.

4.2 LITERATURE REVIEW

The literary review was performed from a structured search in indexed databases. Documents published in scientific journals were revised between the end of 2019 and the end of 2020. A bibliometric analysis of the keywords and countries studying the topics related to this research was then carried out.

Around 274 documents were identified in the literary review. VOSviewer software was used for the bibliometric analysis of these documents (van Eck and Waltman, 2019). Below are the maps and tables based on the systemic review with which the relationships between the study concepts were identified:

- Figure 4.4 and Table 4.1 present the 25 words of 2,301 keywords most studied by the revised authors. The figure shows the network relationship between different words. While the table shows the keyword, its occurrences in the different articles, and the total of links between documents.
- Figure 4.5 and Table 4.2 show the top 25 countries in 84 countries discussed in the documents, which you research on the topics of study. The figure shows the network relationship between different countries. While the table shows the Country, the number of documents produced, the number of subpoenas, and the total number of links between documents.
- Figure 4.6 and Table 4.3 set out the top 25 authors of 1,187 identified authors. The figure shows that the relationship between the authors is isolated. While the table shows authors, the number of documents published by them, the number of subpoenas, and the total number of links between documents.
- Figure 4.7 set out the top 25 author's thermal maps. The figure shows that the relationship between the authors is isolated documents. Also, the map shows the strong research groups.
- Figure 4.8 show the top 25 countries' thermal map. The figure shows that the United States is the strongest research country.

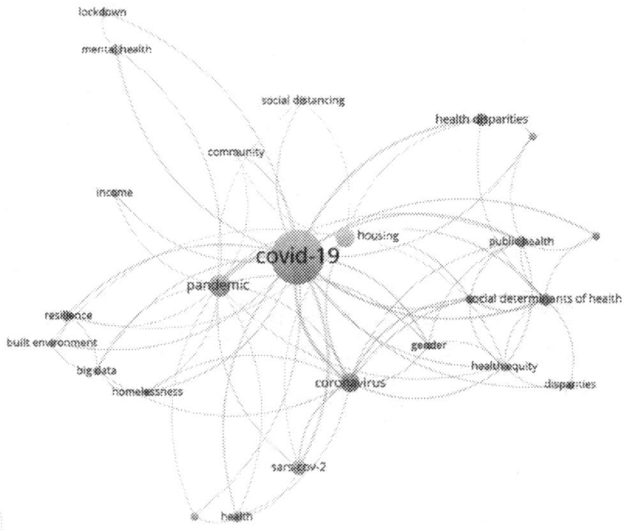

FIGURE 4.4 Map keyword relationships according to literary review.
Source: Own elaboration in VOSviewer.

TABLE 4.1 25 Keywords Most Studied by the Authors Consulted

Id	Keywords	Occurrences	Total Link Strength
1	COVID-19	125	106
2	Pandemic	21	36
3	Coronavirus	16	31
4	Housing	16	31
5	SARS-Cov-2	10	15
6	Health disparities	8	11
7	Social determinants of	7	17
8	Health	6	8
9	Mental health	6	7
10	Public health	6	11
11	Resilience	6	8
12	Health policy	4	12
13	Social distancing	4	5
14	Big data	3	7
15	Built environment	3	6
16	Community	3	5
17	Disparities	3	4
18	Gender	3	8
19	Health equity	3	11
20	Health inequities	3	5
21	Homelessness	3	5
22	Income	3	3
23	Lockdown	3	3
24	Public policy	3	5
25	Quarantine	3	4

Source: Own elaboration in VOSviewer software.

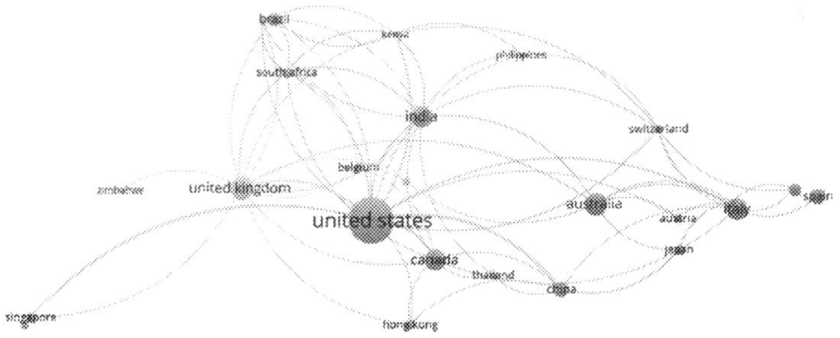

FIGURE 4.5 Map of country relations according to literary revision.
Source: Own elaboration in VOSviewer software.

TABLE 4.2 25 Countries that Study Research Topics the Most

Id	Country	Documents	Citations	Total Link Strength
1	Australia	23	26	9
2	Austria	3	2	3
3	Belgium	4	3	5
4	Brazil	8	60	6
5	Canada	20	49	13
6	China	11	23	11
7	Egypt	2	36	7
8	France	7	5	4
9	Hong Kong	4	0	4
10	India	20	62	17
11	Italy	20	43	9
12	Japan	5	4	4
13	Kenya	2	36	10
14	Netherlands	3	4	3
15	Philippines	2	0	4
16	Singapore	5	10	3
17	South Africa	5	41	12
18	South Korea	3	3	2
19	Spain	11	6	2
20	Switzerland	3	5	8
21	Thailand	3	3	3
22	United Kingdom	24	369	19
23	United States	93	207	30
24	Viet Nam	2	2	4
25	Zimbabwe	2	0	2

Source: Own elaboration in VOSviewer software.

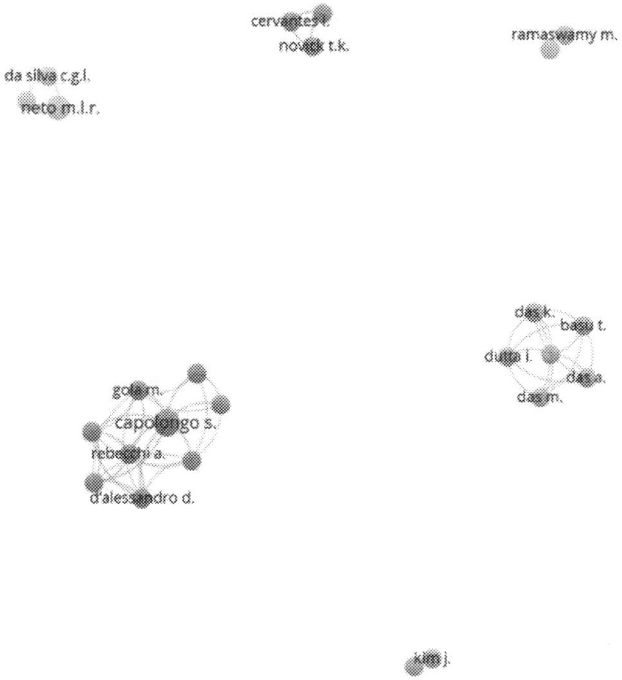

FIGURE 4.6 Map of authors' relationships according to literary revision.
Source: Own elaboration in VOSviewer software.

TABLE 4.3 25 Authors Who Study the Subjects of Research the Most

Id	Author	Documents	Citations	Total Link Strength
1	Appolloni L.	2	18	10
2	Basu T.	2	3	10
3	Brambilla A.	2	5	6
4	Capolongo S.	4	23	16
5	Cervantes L.	2	3	4
6	Da Silva C.G.L.	2	14	4
7	Das A.	2	3	10
8	Das K.	2	3	10
9	Das M.	2	3	10
10	De Souza R.I.	2	14	4
11	Dutta I.	2	3	10
12	D'Alessandro D.	2	18	10

TABLE 4.3 *(Continued)*

Id	Author	Documents	Citations	Total Link Strength
13	Fara G.M.	2	18	10
14	Ghosh S.	2	3	10
15	Gola M.	2	7	8
16	Kim J.	2	0	2
17	Lee S.	2	0	2
18	Morganti A.	2	5	6
19	Neto M.L.R.	3	20	4
20	Novick T.K.	2	3	4
21	Ramaswamy M.	2	0	2
22	Rebecchi A.	2	18	10
23	Rizzolo K.	2	3	4
24	Signorelli C.	2	16	8
25	Wickliffe J.	2	0	2

Source: Own elaboration in VOSviewer software.

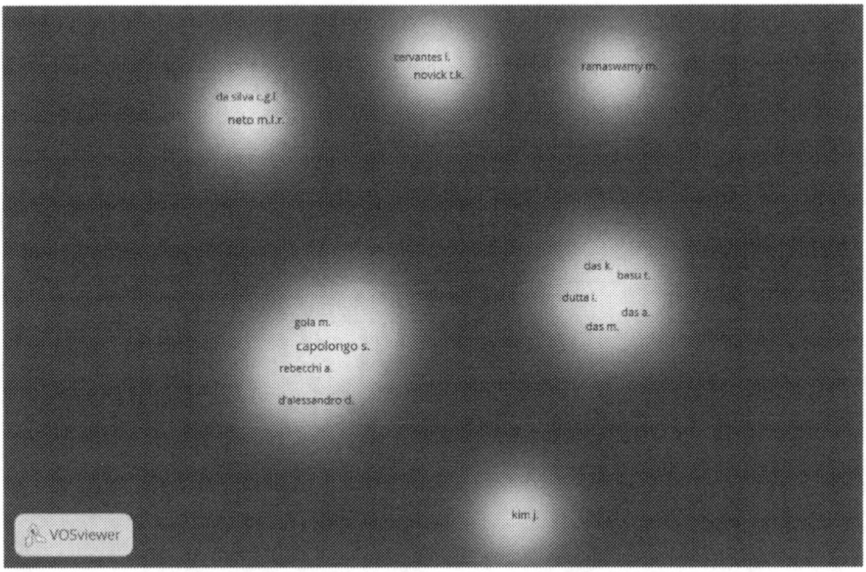

FIGURE 4.7 Thermal map of authors relations according to literary revision.
Source: Own elaboration in VOSviewer software.

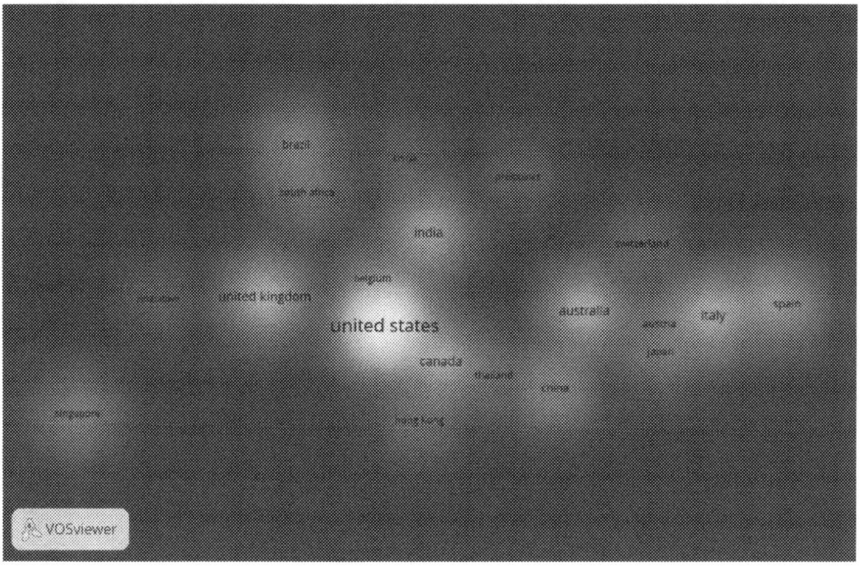

FIGURE 4.8 Thermal map of country relations according to literary revision.
Source: Own elaboration in VOSviewer software.

4.3 CONSTRUCT RESEARCH

The research constructs are presented below according to the literary review. For these three constructs will be developed. These are Affordable Housing, Resilient Design, and Healthy Environments. For later conclusions of this work.

4.3.1 AFFORDABLE HOUSING

The COVID-19 pandemic has transformed social practices during the closure period of the different countries (D'alessandro et al., 2020). Indeed, it observed that the way of working and daily life of people has changed with physical distancing. What causes their social practices to be different? Also, the spaces that make up the buildings are required to be resilient. It is especially true in affordable housing. This need for resilience is related to the interior and exterior spaces that are related to the house (see Figure 4.9).

FIGURE 4.9 Resilience is related to the interior and exterior spaces that are related to the house, 2008.
Source: Photographs of the author.

Therefore, it is necessary to evaluate livability to identify the role that variables such as hygiene and comfort of the home have currently. Then, it is a series of strategies to answer the need for resilience. They are critical replies for dwellings in the face of the pandemic today. These strategies are:

- Green spaces accessible both outside and inside the home (Cubillos-González et al., 2017; D'alessandro et al., 2020).
- Flexibility of living spaces (Cubillos-González, 2006).
- Application of principles of sustainable architecture and Indoor Air Quality (D'alessandro et al., 2020).
- Water consumption and wastewater management (D'alessandro et al., 2020).
- Management of solid urban waste (D'alessandro et al., 2020).
- Home automation and electromagnetic fields (D'alessandro et al., 2020).

- Construction materials and interior finishing (Cubillos-González, 2017; Cubillos-González et al., 2017).

These recommendations can be initial guidelines to control the pandemic as a starting exercise of parameters that can guide the design of housing and particularly that of affordable housing towards resilience. For example, adequate resilience strategies in affordable housing would not make sense if social equity parameters are not taken notice that allows social distancing not to isolate the inhabitants of this type of dwelling (Silva, 2020).

In this sense, it is significant to note that the isolation of the inhabitants of affordable housing affects their well-being and makes, even if the measures of staying at home are advised to reduce infections. It is also true that people must find a balance between living indoors and having access to safe public spaces. The pandemic has generated the need to re-evaluate the affordable housing production model in this context. Also, there are opportunities to propose new forms of habitability that allow its users to have quality (Maalsen et al., 2020).

Likewise, it is essential to identify the role that housing plays in this pandemic since this situation has generated an image that appears that housing offers security that other types of buildings do not guarantee. However, at present, no research shows, from the design point of view, elements of mitigation of the vulnerability that the inhabitants of the dwellings and particularly the low-income inhabitants who live in affordable housing may have. For example, what happens to residents who are at high risk for COVID-19 from serious illnesses? (Mericle et al., 2020).

It is for this reason that housing experts play a decisive role today. So, housing is emerging as a challenge for the coming years. Many of the strategies to respond to COVID-19 are focused on obtaining a safe vaccine that contains the epidemic. Therefore, it is the responsibility of housing specialists to face the crisis (Rogers and Power, 2020). The answers should be re-evaluating the housing production systems and particularly those of affordable housing, that they guarantee good livability and high health security for their users.

In this sense, this pandemic has proven to be a challenge for the city and especially for urban development. Urban approaches with high densification of cities conflict with the pandemic. The increase in density decreases the social distancing required in terms of hygiene (Mahendra and Mittal, 2020).

On the other hand, housing is one of the fundamental components of the urban form (see Figure 4.10). How is this type of urban pattern affected in the face of the new need for more space to distance people to make the city more livable and safer?

FIGURE 4.10 Housing is one of the fundamental components of the urban form.
Source: Rolando Cubillos, Bogotá, Colombia (2019).

Here we return to the issue of quality. These urban forms that large areas of the dwelling. Particularly, affordable housing must guarantee minimum conditions of habitability with internal and external spaces that will warranty the non-spread of viruses and reduce the potential contagion between communities (see Figure 4.11).

Here it is crucial to consider the following (Cao et al., 2020):

- The consumption of more economical housing space implies a greater probability of residential overcrowding.
- Spatial variation in terms of geographic location, local demographics, economic growth, and development.

In this sense, actions like this have a recovery approach after the pandemic. Also, it is essential to analyze the behavior of the market and housing prices, especially in the area of affordable housing (Albuquerque et al., 2020).

Another main point is the relationship between technology and housing. A crucial factor is the traceability of the movements of the inhabitants when they go out to buy supplies. In this context, this type of technology shows social inequalities (Rosenberg et al., 2020). Not all inhabitants are willing to be traced. It implies several rights issues that require in-depth study.

In this sense, digital technologies and data become a new form of metric that arises in response to the pandemic. Two themes emerge here about the notion of home from the perspective of the private: the first digital technology and the second home surveillance (Maalsen and Dowling, 2020). These two variables have implications for the inhabitants of affordable housing and their corresponding privacy. It implies that the workspaces will be into the home as an additional space. Also, it requires consideration in the definition of its design since introducing work at home does not mean de-formalizing its requirements and standards.

On the other hand, there is the concept of a second home. Since the inhabitants who had a second housing option moved from densified centers to rural or low-density areas (see Figure 4.12). These inhabitants Have the advantage over all those who have a single home that is especially true in affordable housing clusters, which are dense (Zoğal et al., 2020). However, the literary review showed that the apparent safety of second homes was false since second homeowners were vectors of transmission of the virus. Therefore, it is necessary to consider actions that mitigate this type of practice to reduce contagion between areas of different densities.

In this regard, it is pertinent to reflect on the different uses of housing that make it mixed, welcoming other uses such as work and tourism (Rubino et al., 2020). What kind of configuration will the home have after the pandemic? Will affordable housing have other uses? Will it remain the same as the old normal? These are pertinent questions. Since there is a need to evaluate the acceptable minimums that affordable housing requires for low-income inhabitants. Will affordable housing guarantee the same security as post-pandemic housing?

In short, there is a great need to propose new guidelines for the design of affordable housing. As can be seen, it tends to have disadvantages in its design concerning the emerging issues that today require a dwelling that responds to the pandemic. Therefore, the need for a healthy and comfortable living space is essential for physical and mental well-being (Tokazhanov et al., 2020). This well-being must not only respond to the requirements of the pandemic but also the needs of sustainability and climate change. It reinforces the idea of the need for resilience in housing in general and affordable housing. It should be that today technology plays a crucial role when designing a home.

FIGURE 4.11 Spaces that will warranty the non-spread of viruses and reduce the potential contagion between communities, 2016.
Source: Photographs of the author.

FIGURE 4.12 Second housing option moved from densified centers to rural or low-density areas, 2007.
Source: Photographs of the author.

Returning to urbanism, and planning, today's city design must overcome the challenge of the pandemic (Ahsan, 2020), guaranteeing a safe habitat, and highlighting that housing has in the configuration of the city. In this sense, the city requires safe public and private spaces. From the perspective of affordable housing, it should tend to reduce its densification (see Figure 4.13). This strategy enhances pandemics due to the reduced spaces and the proximity of the inhabitants.

Another important topic to touch on is the relationship of the pandemic with environmental impacts, an example of which is waste management systems (Ikiz et al., 2021). For instance, preventive isolation has generated the production of waste, making its management, and disposal more complex. This increase is a potential focus of the spread of the pandemic virus and other viruses. This phenomenon is most evident in high-density housing clusters, where affordable housing stands out.

Regarding this issue, it is observed that waste management in high-density homes has had the following changes (Ikiz et al., 2021):

- Changes in the garbage, recycling, and organic flows;
- New health and safety concerns;
- Changes in reuse and reduction practices;
- Changes in the collection of waste and deposit-return bottles;
- Changes in education on waste diversion and reduction.

On the other hand, it observed that public space, mostly green areas, are tending to become critical spaces, due to social distancing and the isolation of homes. Considerable isolation between buildings (see Figure 4.14), together with greater amounts of green areas could be an adequate response to the design of the city in the face of pandemic phenomena (Osborne et al., 2020).

4.3.2 RESILIENT DESIGN

The COVID-19 pandemic is causing a drastic reduction in social and economic practices in cities. In this sense, the design of housing requires identifying barriers and opportunities in the short, medium, and long term. To this, it is necessary to add studies of the housing market in the face of the pandemic (Del Giudice et al., 2020). In this sense, resilient design plays a fundamental role today. Affordable housing cannot see from only the economic dimension. It requires a systemic vision that can compensate for the other aspects.

FIGURE 4.13 Perspective of affordable housing, it should tend to reduce its densification, 2018.
Source: Photographs of the author.

FIGURE 4.14 Social distancing and isolation between buildings, 2018.
Source: Photographs of the author.

The resilient design must be multiscale to study the phenomenon of housing from the urban and in turn from the building (see Figure 4.15). Likewise, the responses to the event of the pandemic are not only located on a hygiene issue but also require understanding that affordable housing must tend towards equality in its services for other types of constructions (Cenecorta, 2020). Therefore, technology and innovation in its design are necessary to respond to the challenges that the pandemic presents today.

Among the variables of resilient design is energy efficiency. A phenomenon like the pandemic increase's energy consumption due to the confinement of the inhabitants of affordable housing. Some studies show the increase in energy consumption in low-income households (Brown et al., 2020) and propose some guidelines in this regard:

- Identify the link between the energy load of the home and health;
- Implement solar energy in buildings;
- Identify the ranges of stress due to energy consumption, as climate change introduces new stress layers that challenge the transition to a clean energy future.

Another point to consider in resilient design is the factor of the age of the inhabitants of affordable housing. Since the age differences represent different vulnerabilities for each type of family. For households of productive age, the pandemic represents a vulnerability in their income, while families that are close to retirement their vulnerabilities are more related to health (Mikolai et al., 2020). However, for low-income families living in affordable housing, this type of vulnerability is more evident, putting family stability at risk.

Indeed, although the health risks are higher in people of retirement age, this does not imply a risk for inhabitants of other ages. In this sense, the designer of the house must orient the design to more ventilated areas, but spaces of isolation and safe encounter within the house. In this sense, it is necessary to evaluate the urban form both in its public and private spaces. In this sense, some authors propose principles to improve resilience through the design of the city in response to the pandemic (Lak et al., 2020):

- The urban form in three scales of housing, neighborhoods/public spaces, and cities;
- The concept of a resilient urban structure from new perspectives focused on physical and non-physical aspects;

- The physical form of resilience access infrastructure, land use, and natural environment factors.

FIGURE 4.15 Resilient design must be multiscale study of housing from the urban and in turn from the building, 2019.
Source: Photographs of the author.

Therefore, it is crucial to combine resilient design actions at the urban, architectural, and interior space levels. That is multiscale actions. It could produce positive impacts on future resilience designs, plans, and policies within the housing and built environment (Keenan, 2020).

It is critical to consider the physical and mental health of the inhabitants of affordable housing since the reduction of social interaction has adverse effects on the population (see Figure 4.16).

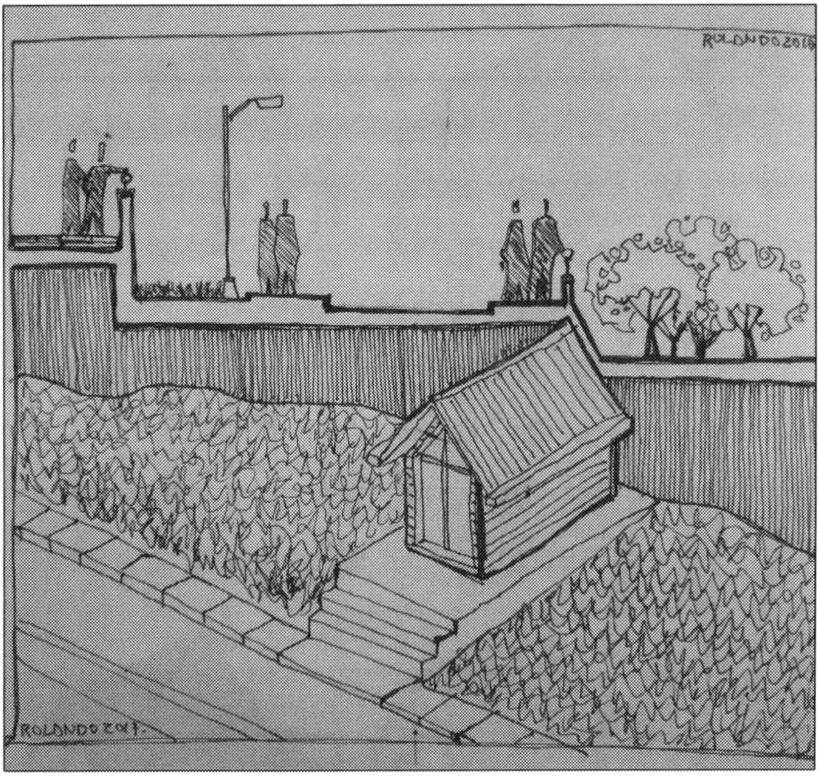

FIGURE 4.16 Physical and mental health of the inhabitants of affordable housing, 2017.
Source: Photographs of the author.

These effects can increase the consequences of deteriorating health for the inhabitants of affordable housing. Some studies show the importance of a multiscale interconnected urban form (Sliwa and Yacoub, 2020; Xie et al., 2020). In this sense, the authors propose:

- Emphasizes the role of urban parks from the perspective of the construction environment;
- Large open spaces outdoors to provide residents with a place for safe social interaction and activities;

- Buffer areas to maintain positive health and quality of life. According to some authors (Sharifi and Khavarian-Garmsir, 2020), today, the design of cities and particularly the resilient design of cities is related to:
  - Environmental quality;
  - Socio-economic impacts;
  - Management and governance;
  - Transport and urban design.

In this context, actions to reduce environmental impacts are a significant issue to mitigate the possibility of possible infections and new pandemics. It is an excellent opportunity for planners to take transformative action to create more resilient and sustainable cities (see Figure 4.17).

Here we return to energy efficiency within urban planning policies (Brosemer et al., 2020). Affordable housing areas must maintain energy sovereignty. It would allow these areas not to enter an energy vulnerability, which would put this type of population at greater risk. Alternative energy plays an important role here, where fair and equitable access to essential energy services is guaranteed.

In this sense, Capolongo et al. (2020) propose immediate actions that change social practices and train the city and housing process according to these actions:

- Schedule the flexibility of city hours;
- Plan a smart and sustainable mobility network;
- Define a neighborhood service plan;
- Develop digitization of the urban context, promoting smart communities;
- Rethinking accessibility to places of culture and tourism. Actions in the medium and long term;
- Design the interior flexibility of domestic living spaces;
- Rethink building typologies, promoting the presence of semi-private or collective spaces;
- Renew the network of basic care services;
- Integrate existing environmental emergency plans with those related to health emergencies;
- Improve stakeholder awareness of factors that affect public health in cities.

FIGURE 4.17 An excellent opportunity for planners to take transformative action to create more resilient and sustainable cities, 2019.
Source: Photographs of the author.

Cvetković et al. (2021) identify an increase in the consumption of water and energy in the residential sectors. This implies that designers should consider four possible design scenarios to face the increase in consumption:

- **S1:** Normal state.
- **S2:** Mild protective measures.
- **S3:** Semi-quarantine measures.
- **S4:** Full quarantine.

The construction of these scenarios would allow designers to visualize the behavior of people about their energy consumption depending on the situation. It is also to take into account that the COVID-19 pandemic is not distributed evenly (Nguyen et al., 2020). This implies that designers must identify their design strategies based on risks and available resources.

This is where resilient design processes come in (see Figure 4.18), acting as mitigators or reducers of pandemic risk. In this sense, technology can contribute from the interactivity of the images in real-time of the city. This would allow remote control of urban spaces to identify the proper use of public space and social distancing. Computer vision makes possible studies of the effects of the built environment on the risk of COVID-19, to inform planners of decision making in the local area (Nguyen et al., 2020).

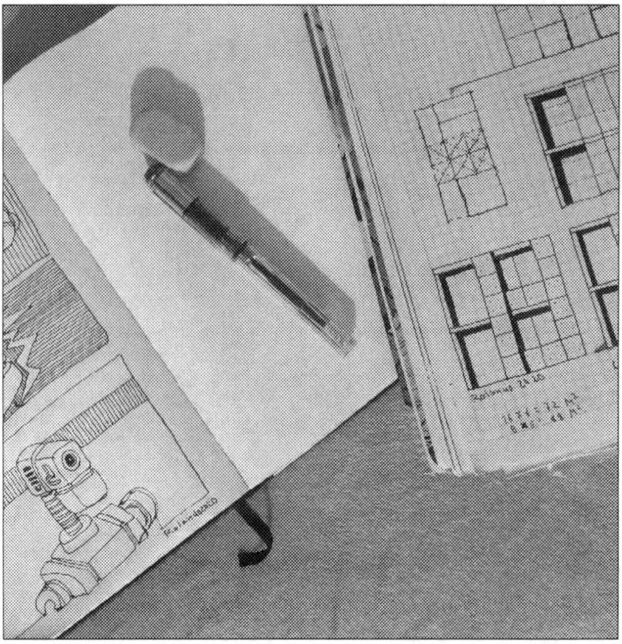

FIGURE 4.18 The resilient design processes come in, 2020.
Source: Photographs of the author.

4.3.3 HEALTHY ENVIRONMENTS

Healthy environments are characterized by guaranteeing safety in all aspects of the different components of the inhabited space. For example, a healthy environment in the case of homes, such as the kitchen, should guarantee the preservation of food and its perfect storage. An example of this is the control of humidity in food (Wu et al., 2020).

Therefore, the effects of the pandemic are not limited to simple control of contagion of the virus. Actions to maintain the survival and viability of affordable housing are also important. In this sense, it is important to identify possible sources of diseases both in the internal spaces of the home, as well as in the spaces outside it and in public spaces.

The health effects of the pandemic are not limited to mere mitigation. These extend to other dimensions of daily life, which must be considered in the affordable housing design process. Furthermore, when the population has different perspectives on this phenomenon (Antipova, 2020). The house reflects its inhabitants, and a safe house inspires its inhabitants to self-care. Likewise, the designer must understand that the previous flexibility processes must be oriented through the same spatial configuration (see Figure 4.19). This would prevent adaptation processes from moving away from minimum standards of physical and mental health.

This situation forces the designer to have the user in their design. It is important to identify the needs of affordable home users and translate these needs into actions that include resilience and mitigation of potential future pandemics. Space configuration in affordable housing is intimately linked to phenomena such as domestic violence (Sacco et al., 2020). In this sense factors such as stress, long periods in minimum spaces Increase the risk of domestic violence.

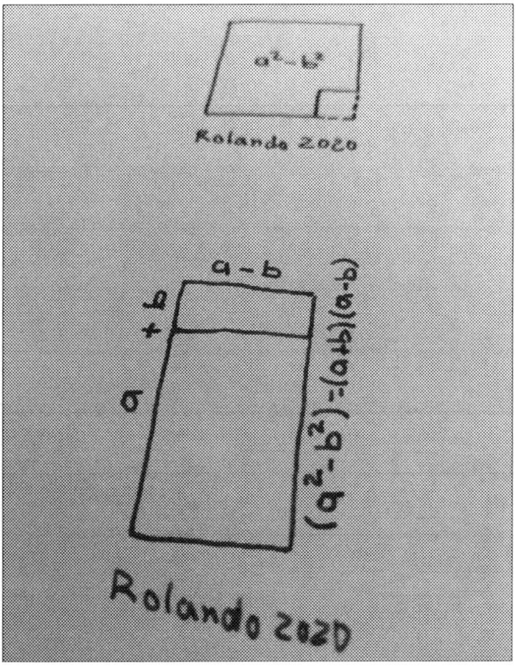

FIGURE 4.19 The resilient design processes come in 2020.
Source: Photographs of the author.

Besides, the flexibility mentioned above suggests that designers explore the ability to adapt a home to short, medium, and long periods inhabiting the same space. What spatial capacity can be supplied by designers to avoid monotony, depression, and other diseases. How could this spatial design ensure this flexibility of space that dynamism to the habitability of housing and especially affordable housing?

In this sense, authors such as Amerio et al. (2020) said that housing is a determinant of health. Here the size of the house plays an important role. How to reconcile current minimum housing standards, with the new demand for greater spaces to reduce the risks of contagion. Is it possible for the pandemic to reassess these standards? Is it possible that health standards can be imposed on the economic standards of minimum space?

The arguments set out here open a door to the possibility of the better spatial quality of affordable housing. So that the minimums are maximum with the physical and mental health needs of the inhabitants of affordable housing (see Figure 4.20). In this respect, the literary review shows that people with different disorders and diseases may die in spaces of unsuitable sizes (MacKinnon et al., 2020).

FIGURE 4.20 Physical and mental health needs of the inhabitants of affordable housing, 2020.
Source: Photographs of the author.

It is also important to understand how affordable housing meets the demand of low-income people. Homeless people or people living in precarious housing are high-risk agents in pandemic terms. Ensuring them adequate affordable housing is critical (Ralli et al., 2020). Preventing the increase of COVID-19 cases in low-income residents requires the application of physical spaces that guarantee habitability to members of such communities.

In this respect, healthy spaces are the fundamental axis of good affordable housing design. Since these spaces allow to guarantee the mental and physical health of users. For designers, it is required to change the conventional design to that of an evidence-based design process (see Figure 4.21). That you allow access to scientific data for an adequate response to affordable housing users.

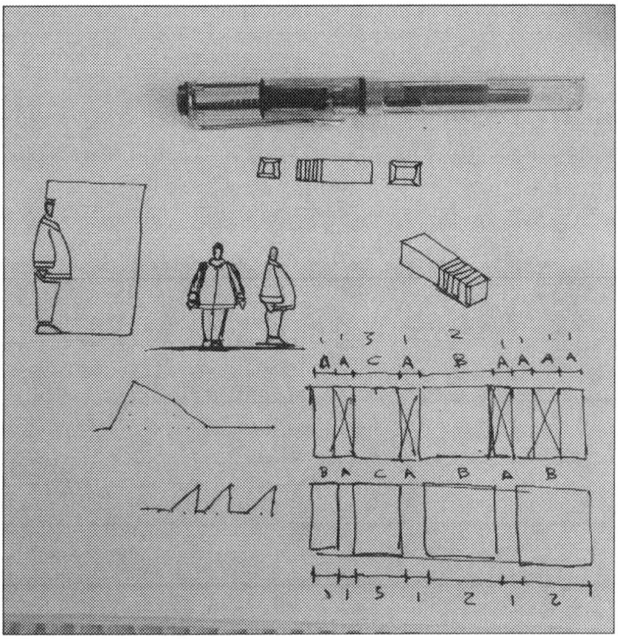

FIGURE 4.21 To change the conventional design to that of an evidence-based design process, 2020.
Source: Photographs of the author.

Now authors such as Fazzini et al. (2020) explain the climate factors that influence the spread of the pandemic. Indeed, variables such as temperature and humidity play an important role in mitigating contagion conditions. Especially when it comes to respiratory diseases, in this sense, the environmental comfort control of affordable housing is very important. Also, the choice of materials is essential when controlling a future pandemic. Identifying the time of stay and subsistence in the materials allows the designer, to control the possible outbreaks of the virus.

Besides, an adequate identification of ventilation strategies in the house allows a successful control of the concentration of the virus in the spaces of the house. Internal air circulation is essential for this purpose. Authors such as Domínguez-Amarillo et al. (2020) place special emphasis on this issue. For housing designers, it is important to consider indoor and outdoor temperatures, variation in internal CO_2 production, particulate entry into housing and concentration of volatile gases.

Although with the pandemic the different populations reduced their exposure to the pollution of the city. Internal housing pollution increased considerably (Domínguez-Amarillo et al., 2020), which involves a series of considerations of ventilation in affordable housing (see Figure 4.22). Within the design, indoor air quality should be considered. Above all, ventilation processes should be provided when a dwelling enters in case of total or partial quarantine to protect residents from respiratory diseases.

FIGURE 4.22 A series of consideration of ventilation in affordable housing, 2013.
Source: Photographs of the author.

Authors such as DiMaggio et al. (2020) argue that it is essential to identify demographic characteristics. Since focusing on individual Clinical Factors is not enough. Since areas with large residential proportions have a significantly higher risk according to environmental characteristics and pre-existing conditions in the population.

However, it is essential to identify how housing helps not only to protect the inhabitants, but it must also to allow an adequate recovery of potential patients at home (Clapp et al., 2020). How to ensure control of the spread of a resident in a dwelling against other people who live with it?

In terms of design, this makes affordable housing technically designed in terms of health (see Figure 4.23). In this sense, authors like Saez et al. (2020) claim the risk of infection and death from COVID-19 could be associated with heterogeneous distribution at the level of small areas and environmental factors. It is important to introduce control elements into the house that reduce exposure to carbon dioxide (CO_2), large amounts of particulate matter that increases the risk of acquiring not only the virus but respiratory diseases.

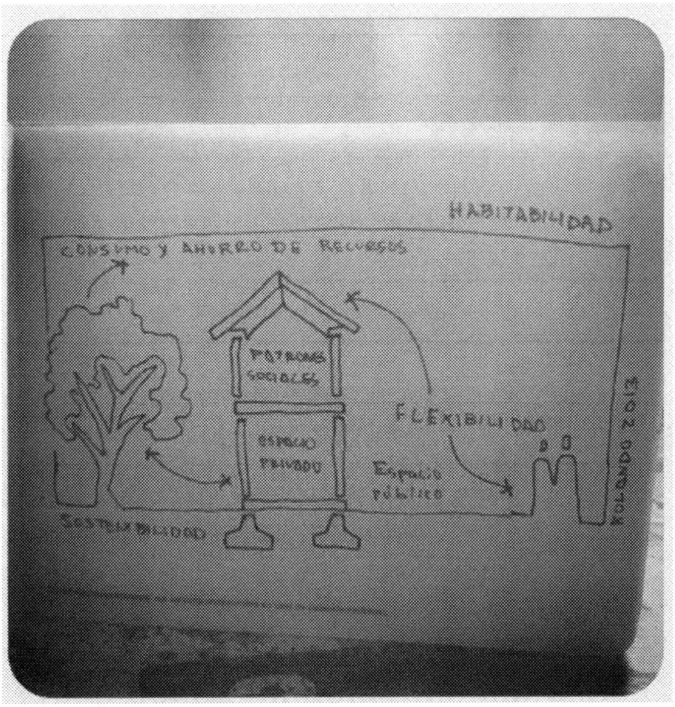

FIGURE 4.23 Affordable housing design in terms of health, 2013.
Source: Photographs of the author.

Combined with these factors, it is also necessary to study, the adequacy of spaces for recreation and sport that are typical of affordable housing. For example, some studies (Pombo et al., 2020) pose the need for their daily routines to confinement for the inhabitants of the dwellings. These factors involve evaluating on the designer's part, family size, different ages, physical abilities, and limitations.

It is identified that it is important to learn from past lessons in terms of environmental health and sustainability (see Figure 4.24). Resilience is that ability to maintain certain capabilities despite events affecting our environmental, physical, social systems, etc. (Moustafa, 2020). In this sense, we return to the need for design support in technology. Digital production is essential in this context.

FIGURE 4.24 Past lessons in terms of environmental health and sustainability, 2013.
Source: Photographs of the author.

The pandemic provides an opportunity to strengthen digitization processes in the construction sector. Giving the possibility to provide affordable housing using 3D printing to build houses quickly. Implementing such technologies can help solve affordable housing demand crises more efficiently. About traditional production that is limited in responding adequately to the needs of phenomena such as the COVID-19 pandemic.

4.4 PANDEMIC, HOUSING, AND URBAN DESIGN

Authors such as Sharma et al. (2020) put up a series of guidelines that are relevant to be included in an affordable housing design process:

- Fear of getting COVID-19;
- Disruption of the employment situation;
- Financial difficulties;
- Aggravation of food insecurity.

According to Adlakha and Sallis (2020), friendly, and walkable neighborhoods are suitable for environmental sustainability and good economic performance. On the other hand, other authors are concerned about the risks posed by high-density urban residential areas, particularly with events such as the pandemic.

However, it is noted that dense mixed-use neighborhoods and recreational areas can reduce the risk of contagious diseases. Indeed, Adlakha and Sallis (2020) propose the following strategies to be implemented in residential urban areas after the pandemic (see Figure 4.25):

- Increase in parks, trails, and open spaces;
- Increase in mixed uses in residential areas;
- Increase in recreational facilities and routes for cyclists.

The author concludes that the reduction of differences in the health of the inhabitants of high-density urban areas creates communities favorable to physical activity and reduces the chances of contagion in the population.

On the other hand, Aleksić et al. (2016) explain that climate change is one of the most relevant challenges facing the 21st century. Indeed, society faces two ways to solve this crucial challenge. First, housing design strategies that contribute to the reduction of climate change. Second, the importance is to implement a resilient building process. These strategies types can be an adequate response to the current pandemic situation.

FIGURE 4.25 Parks, trails, and open spaces, 2019.
Source: Photographs of the author.

Since they offer new opportunities to include health strategies that reduce the effects of the pandemic in homes and allow a new dynamic in the design of residential and urban areas. That may identify that the need for resilience in the face of the pandemic is only one more element of the mitigation process that must introduce in the face of climate change.

Cole (2020) explains how building designers play an essential role in the responses after the COVID-19 pandemic. Adaptation to different situations caused by the effects of climate change and the pandemic largely depends on the actions taken by building users in their daily lives (see Figure 4.26). Therefore, the design of buildings must respond to these needs.

It is essential to change the focus from an individual design that responds to environmental impact, to a building design approach as part of a broader and more systemic urban system.

FIGURE 4.26 Adaptation to different situations caused by the effects of climate change and the pandemic, 2020.
Source: Photographs of the author.

Cole (2020) proposes three themes for the design of buildings:

- First, strategies are based on science and not on political or economic priorities;
- Second, Identify the need to understand the relationship between crises and public responses;
- Third, Identify the vulnerabilities of the different population groups.

In this sense, developing a clearer vision of the impacts of the pandemic will be a long-term project for housing scholars (Druta et al., 2020). The pandemic has affected the concept of housing in its economic, social, and environmental dimensions. By the way, the first months of 2020 showed considerable dynamism in mortgage loans in Europe, while house prices continued to rise in many contexts. For example, in the Netherlands, the values of these three dimensions have been reconsidered against the concept of housing.

In the economic dimension, housing has been affected in its market. Although several governments have provided financial support to the sector, it is possible that in the future, when such support ends, the market will show stagnation.

In the social dimension, the pandemic has highlighted the problems of affordability and low quality of some homes. This is true with the most vulnerable sectors of the population. Likewise, urban densification has entered a crisis and requires re-evaluating its implementation in cities (see Figure 4.27). Finally, in the Environmental dimension, the argument of designing a sustainable and resilient home in the face of climate change mitigation is reinforced.

All these concerns pose a challenge to designers, planners, economists, and politicians. Since it is necessary to reconsider current housing policies worldwide, a single argument must be replaced by a multi-criteria vision that allows systemic responses that respond to the complexity.

On the other hand, Gilderbloom and Meares (2020) propose that residential urban areas and public spaces will have an increase in their value since these spaces types, as they have greater possibilities of outdoor traffic, reduce pollution rates, and circumstances of contagion due to the pandemic.

FIGURE 4.27 Urban densification has entered a crisis and requires a revaluation of its implementation in cities, 2019.
Source: Photographs of the author.

Another critical point is the study of the safety of public spaces in residential urban areas (He et al., 2020). Since this variable is directly related to quality and the built environment. The perception that the inhabitants have of their different public spaces is of great value. The pandemic has left a vast cost in the role of public space as an articulator of areas that guarantee social distancing (see Figure 4.28).

Likewise, authors such as Ikiz et al. (2021) showed how waste disposal patterns have considerably transformed during the pandemic. This is especially significant in multi-family homes. For example, this is the case of the disposal patterns of the inhabitants of Toronto's multiple dwelling buildings in Canada. Where the following evidenced during the partial lockdown (Ikiz et al., 2021):

- Changes in garbage, recycling, and organic flows;
- New health and safety concerns;
- Changes in reuse and reduction practices;
- Changes in the collection of appropriate waste and return of deposit bottles;
- Changes in waste diversion and reduction in education.

FIGURE 4.28 The role of public space as an articulator of areas that guarantee social distancing, 2019.
Source: Photographs of the author.

4.5 AFFORDABLE HOUSING, RESILIENT DESIGN, AND HEALTHY ENVIRONMENTS FRAMEWORK

A synthesis of the theoretical framework shows below that allows identifying the most important strategies of what has been exposed in this chapter (see Table 4.4).

TABLE 4.4 Theoretical Framework

Affordable Housing	Resilient Design	Healthy Environments
Control elements into the house that reduce exposure to the virus.	Identify the link between the energy load of the home and health.	The urban form in three scales of housing, neighborhoods, public spaces.
Green spaces accessible both outside and inside the home.	Implement solar energy in buildings.	The urban form in three scales of housing, neighborhoods, public spaces.
Flexibility of living spaces.	Identify the ranges of stress due to energy consumption.	The physical form of resilience access infrastructure, land use, and natural environment factors.
Spatial variation in terms of geographic location, local demographics, economic growth, and development.	Application of principles of sustainable architecture and indoor air quality.	Environmental quality.
Water consumption and wastewater management.	Transport and urban design.	Socio-economic impacts.
Define a neighborhood service plan.	Rethink building typologies, promoting the presence of semi-private or collective spaces.	Management of solid urban waste.
Plan a smart and sustainable mobility network.	Home automation and electromagnetic fields.	Schedule the flexibility of city hours.
Increase in recreational facilities and routes for cyclists.	Construction materials and interior finishing.	Develop digitization of the urban context, promoting smart communities.
Increase in mixed uses in residential areas.	Increase in parks, trails, and open spaces.	Renew the network of basic care services.

Source: Own elaboration.

Affordable housing must be designed to reduce exposure to disease after the pandemic. For this, it is essential to guarantee broad and flexible spaces that allow minimal social distancing if necessary. In these terms, the design of affordable housing does not refer only to the mass production housings, but to the resilient design of housing, neighborhoods, and communities (see Figures 4.29 and 4.30). Therefore, the affordable housing design must be next to urban design and master plans that guarantee the optimal development of residential urban areas.

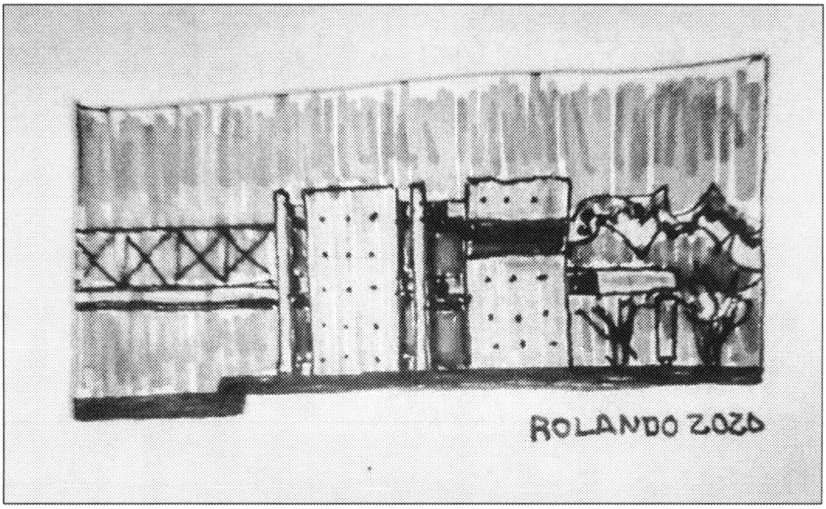

FIGURE 4.29 Resilient housing, neighborhoods, and communities, 2020.
Source: Photographs of the author.

FIGURE 4.30 Resilient housing, neighborhoods, and communities, 2020.
Source: Photographs of the author.

In this sense, public space, green areas, urban services, and road infrastructure that guarantee enough affordable housing in the face of an event such as a pandemic (see Figures 4.31 and 4.32).

FIGURE 4.31 Public space, green areas, Urban services, 2020.
Source: Photographs of the author.

FIGURE 4.32 Public space, green areas, Urban services, 2020.
Source: Photographs of the author.

Urban densification processes increase the risks of contagion. But at the same time, it offers an opportunity to consolidate a good health infrastructure close to residential urban areas. In this case, mixed uses played an important role in housing design after the pandemic.

On the other hand, the strengthening of resilient design strategies for the design of affordable housing makes it possible to strengthen the adaptation of energy efficiency systems and renewable energies in this type of building. In this case, the study of consumption baselines is crucial. In other words, user behavior is identified in times of pandemic.

Resilient design requires new types of affordable housing. Where factors such as quality, and health must be above factors such as minimum production costs. One more critical point to include is the introduction of automated systems in affordable housing. Automation must be implemented in the production processes and in the vital systems of the home to guarantee an acceptable healthy environment within it.

Resilient design is multiscale, which allows the identification of different relationships between the home, the neighborhood, and the public space. Healthy spaces are linked to urban design responses related to affordable housing because of these relationships (see Figure 4.33). The physical form that determines the resilient design allows a better environmental quality. Also, designers and planners can determine the socio-economic impacts of their interventions.

FIGURE 4.33 Healthy spaces are linked to urban design, 2020.
Source: Photographs of the author.

Finally, another point is that of waste management, a crucial component in the design of healthy spaces. The digitization of the waste management process in residential urban areas would allow better control of waste disposal and the identification of baselines of the behavior of the inhabitants to mitigate future contagions.

4.6 CONCLUSIONS

COVID-19 pandemic is the most important event of the last year. This phenomenon demonstrated several limitations and in turn, opportunities to improve the design of affordable housing worldwide. Since it is an opportunity for improvement in the aspects that ensure that a home is healthy and safe against phenomena such as those of a pandemic (see Figure 4.34).

FIGURE 4.34 A home is healthy and safe against phenomena such as those of a pandemic, 2020.
Source: Photographs of the author.

Indeed, these changes are closely linked to design processes, standards, and sustainability. This is especially true in the case of accessible housing. Because of its design characteristics, it is more vulnerable to such pandemic phenomena.

Three key factors have been identified when making changes to the design of the. accessible homes, these are (1) flexibility, (2) quality of the building, and (3) quality of life. Another important point is the introduction of automated technologies that reduce and control the contact of agents outside of affordable housing. To develop adequate housing design guidelines important to consider three constructs, these are: affordable housing, resilient design and healthy environments.

Regarding affordable housing should start from the design of new social practices that are supported in the satisfaction of resilience for extreme events. In this sense, it is essential to understand habitability, related to variables such as hygiene and comfort. The design must be oriented to accessible green spaces that are internal and external to the house. Flexibility plays an important role in designing affordable post-pandemic housing (see Figure 4.35).

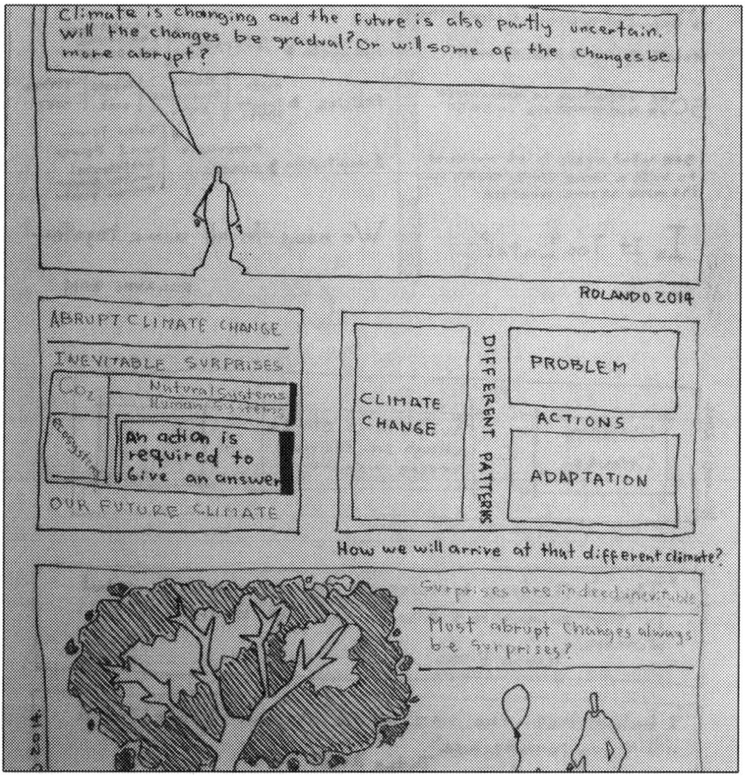

FIGURE 4.35 Flexibility plays an important role in designing affordable post-pandemic housing, 2020.
Source: Photographs of the author.

Since, the adaptability of the internal spaces of the house requires to give a generous number of possibilities so that the inhabitants can live long seasons within the house, without having to leave. It is also important to consider within the design variables, energy, and water consumption of the house.

In addition, it is important to note that an adequate strategy of resilience in affordable housing would not make sense if social equity parameters that allow social stating to not isolated the inhabitants of this type of housing are not regarded. It is also true that people must find a balance between inhabiting the interior space and having access to safe public spaces.

In this context, housing plays an important role today. Because housing design must be aimed at re-evaluating affordable housing production systems so that they ensure good habitability and high health security for its users. On the other hand, housing is one of the fundamental components of the urban form. Such buildings should be projected concerning spatial variation in terms of geographical location, local demographics, economic growth, and development.

In this sense, digital technologies become a new form of metric that emerges in response to the pandemic. Since it involves introducing the workspace into the house as an additional space. Which requires considerations in defining your design, since introducing work at home does not mean de formalizing your requirements and standards. In short, there is a great need to propose new guidelines for the design of affordable housing.

About resilient design, housing design requires identifying barriers and opportunities in the short, medium, and long term. Since affordable housing can no longer be seen from just the economic dimension. It requires a strong systemic vision that can compensate for dimensions other than economic ones (see Figure 4.36).

The resilient design must be multi-scalding, i.e., study the housing phenomenon from the urban and in turn from the building. Therefore, technology, and innovation in its design are necessary to respond to the challenges presented today by the pandemic.

Among the variables of resilient design is energy efficiency. A phenomenon such as the pandemic increases energy consumption due to the confinement of affordable housing dwellers. Another point to consider in resilient design is the age factor for affordable housing dwellers. Since age differences represent different vulnerabilities for each type of family.

Indeed, although health risks are higher in people of retirement age. This does not pose a risk to people of other ages. Therefore, it is important to combine resilient design actions at the urban, architectural, and interior space level. It is important to consider the physical and mental health of affordable housing residents, as reducing social interaction has adverse effects on the population.

FIGURE 4.36 A strong systemic vision that can compensate for dimensions other than economic ones, 2020.
Source: Photographs of the author.

In this context, it is important to reduce environmental impacts that mitigate the possibility of possible contagions and new pandemics. It is therefore crucial that affordable housing areas maintain energy sovereignty. This would allow these areas not to enter an energy vulnerability, to put this type of population at greater risk.

That is where resilient design processes acted as mitigators or pandemic risk reducers come in. In this sense, technology can contribute to the interactivity of the real-time images of the city (see Figure 4.37). Regarding healthy environments, these are characterized by ensuring safety in all aspects of the different components of inhabited space. Therefore, the effects of the pandemic are not reduced to a simple contagion control of the virus. Actions to maintain the survival and habitability of affordable housing are also important.

This situation forces the designer to have the user in their design. It is important to identify the needs of affordable home users and translate these needs into actions that include resilience and mitigation of potential future pandemics.

FIGURE 4.37 Technology can contribute to the interactivity of the real-time, 2020.
Source: Photographs of the author.

Moreover, the flexibility mentioned above suggests that designers explore the ability to adapt a home to short, medium, and long periods inhabiting the same space. Because housing is a determinant of health. Here the size of the house plays an important role.

In this respect, people with different diseases may die in spaces of unsuitable sizes. It is also important to understand how affordable housing meets the demand of low-income people. Preventing the increase of COVID-19 cases in low-income residents requires the application of physical spaces that guarantee habitability to members of such communities (see Figure 4.38).

FIGURE 4.38 Low-income residents require the application of physical spaces that guarantee habitability to members of such communities, 2020.
Source: Photographs of the author.

In this respect, healthy spaces are the fundamental axis of good affordable housing design. For designers, it is required to change the conventional design to that of an evidence-based design process.

Now the climate factors that influence the spread of the pandemic. Indeed, variables such as temperature and humidity play an important role in mitigating contagion conditions. So, the choice of materials is essential to control future pandemics.

Besides, an adequate identification of ventilation strategies in the house allows a successful control of the virus in the spaces of the house. For housing designers, it is important to consider indoor and outdoor temperatures and air quality.

Although with the pandemic, the different populations reduced their exposure to the pollution of the city. Internal housing pollution increased considerably. Above all, ventilation processes should be provided when a dwelling enters in case of total or partial lockdown to protect residents from respiratory diseases.

Since areas with large residential proportions have a significantly higher risk according to environmental characteristics and pre-existing conditions in the population.

However, it is essential to identify how housing helps not only to protect the inhabitants, but it must also allows an adequate recovery of potential patients at home. In terms of design, this makes affordable housing technically designed in terms of health.

Combined with these factors, it is also necessary to study, the adequacy of spaces for recreation and sport that are typical of affordable housing. It is identified that it is important to learn from past lessons in terms of environmental health and sustainability (see Figures 4.39 and 4.40).

FIGURE 4.39 Healthy sport spaces, 2019.
Source: Photographs of the author.

FIGURE 4.40 Healthy sport spaces, 2019.
Source: Photographs of the author.

In this sense, we return to the need for design support in technology and digital production that Aare essential in this context (see Figure 4.41). For example, to provide affordable housing using 3D printing to build houses quickly. Also, about traditional production that is limited in responding adequately to the needs of phenomena such as the COVID-19 pandemic.

FIGURE 4.41 The need for design support in technology and digital production that are essential in this context, 2020.
Source: Photographs of the author.

On the other hand, friendly and walkable neighborhoods are suitable for environmental sustainability and good economic performance. However, dense mixed-use neighborhoods and recreational areas can reduce the risk of contagious diseases. Indeed, the reduction of differences in the health of the inhabitants of high-density urban areas creates communities favorable to physical activity and reduces the chances of contagion in the population.

Society faces two ways to solve this crucial challenge. First, housing design strategies that contribute to the reduction of climate change. Second, the importance is to implement a resilient building process.

Those strategies offer new opportunities to include health strategies that reduce the effects of the pandemic in homes and allow a new dynamic in the design of residential and urban areas (see Figure 4.42). That may identify the need for resilience in the face of the pandemic.

FIGURE 4.42 The pandemic in homes and allow a new dynamic in the design of residential and urban areas, 2020.
Source: Photographs of the author.

Building designers play an essential role in the responses after the COVID-19 pandemic. The Adaptation strategies to different situations are caused by the effects of building users in their daily lives (see Figure 4.43).

It is essential to change the focus from an individual design that responds to environmental impact, to a building design approach as part of a broader and more systemic urban system. In this sense, developing a clearer vision of the impacts of the pandemic will be a long-term project for housing scholars.

FIGURE 4.43 The adaptation strategies to different situations are caused by the effects of building users in their daily lives, 2020.
Source: Photographs of the author.

Also, residential urban areas and public spaces will have an increase in their value because reduce pollution rates and circumstances of contagion due to the pandemic. Another critical point is the study of the safety of public spaces in residential urban areas. The pandemic has left a vast cost in the role of public space as an articulator of areas that guarantee social distancing. For this, it is essential to guarantee broad and flexible spaces.

Resilient design strategies for the design of affordable housing make it possible to strengthen the adaptation (see Figure 4.44). It requires new types of affordable housing. Indeed, Automation must be implemented in the housing production processes and guarantee an acceptable healthy environment within it. The physical form that determines the resilient design allows for a better environmental quality. Also, designers, and planners can determine the socio-economic impacts of their interventions.

Finally, waste management is a crucial component in the design of healthy spaces. The digitization of the waste management process in residential urban areas would allow better control of waste disposal to mitigate future contagions.

Finally, the arguments set out here open a door to the possibility of the better spatial quality of affordable housing (see Figure 4.45). So that the minimums are maximum about the physical and mental health needs of the inhabitants of affordable housing.

FIGURE 4.44 The design of affordable housing makes it possible to strengthen the adaptation, 2019.
Source: Photographs of the author.

FIGURE 4.45 Better spatial quality of affordable housing 2019.
Source: Photographs of the author.

KEYWORDS

- **COVID-19 pandemic**
- **digital technologies**
- **literary review**
- **pandemic phenomena**
- **phenomenon**

REFERENCES

Adlakha, D., & Sallis, J. F., (2020). Activity-friendly neighborhoods can benefit non-communicable and infectious diseases. *Cities & Health, v4*, 1–5. https://doi.org/10.1080/23748834.2020.1783479.

Ahsan, M. M., (2020). Strategic decisions on urban built environment to pandemics in Turkey: Lessons from COVID-19. *Journal of Urban Management*, *9*(3), 281–285. https://doi.org/10.1016/j.jum.2020.07.001

Albuquerque, B., Iseringhausen, M., & Opitz, F., (2020). Monetary policy and US housing expansions: The case of time-varying supply elasticities. *Economics Letters*, *195*, 109471. https://doi.org/10.1016/j.econlet.2020.109471.

Aleksić, J., Kosanović, S., Tomanović, D., Grbić, M., & Murgul, V., (2016). Housing and climate change-related disasters: a study on architectural typology and practice. *Procedia Engineering*, *165*, 869–875. https://doi.org/10.1016/j.proeng.2016.11.786.

Amerio, A., Brambilla, A., Morganti, A., Aguglia, A., Bianchi, D., Santi, F., Costantini, L., et al., (2020). COVID-19 lockdown: Housing Built environment' effects on mental health. *International Journal of Environmental Research and Public Health Article*, *17*, 5973.

Antipova, T., (2020). Coronavirus pandemic as black swan event. *ICIS 2020: Integrated Science in Digital Age*, *2020*, 356–366.

Brosemer, K., Schelly, C., Gagnon, V., Arola, K. L., Pearce, J. M., Bessette, D., & Schmitt, O. L., (2020). The energy crises revealed by COVID: Intersections of indigeneity, inequity, and health. *Energy Research and Social Science*, *68*, 101661. https://doi.org/10.1016/j.erss.2020.101661.

Brown, M. A., Soni, A., Doshi, A. D., & King, C., (2020). The persistence of high energy burden: Results of a bibliometric analysis. *Energy Research and Social Science*, *70*, 101756. https://doi.org/10.1016/j.erss.2020.101756.

Cao, Y., Liu, R., Qi, W., & Wen, J., (2020). Spatial heterogeneity of housing space consumption in urban China: Locals vs. inter-and intra-provincial migrants. *Sustainability (Switzerland)*, *12*(12). https://doi.org/10.3390/su12125206.

Capolongo, S., Rebecchi, A., Buffoli, M., Appolloni, L., Signorelli, C., Fara, G. M., & D'Alessandro, D., (2020). COVID-19 and cities: From urban health strategies to the pandemic challenge. a decalogue of public health opportunities. *Acta Biomedica*, *91*(2), 13–22. https://doi.org/10.23750/abm.v91i2.9515.

Cenecorta, A. X. I., (2020). The city we would like after COVID-19 | La ciudad que quisiéramos después de COVID-19. *Architecture, City, and Environment*, *15*(43), 1–23.

Clapp, J., Calvo-Friedman, A., Cameron, S., Kramer, N., Kumar, S. L., Foote, E., Lupi, J., et al., (2020). The COVID-19 shadow pandemic: Meeting social needs for a city in lockdown. *Health Affairs (Project Hope), 39*(9), 1592–1596. https://doi.org/10.1377/hlthaff.2020.00928.

Cole, R. J., (2020). *Navigating Climate Change: Rethinking the Role of Buildings*, 1–25.

Cubillos-González, R. A., (2006). Social Housing and flexibility: Bogota´s studi case. ¿Why the inhabitants transform their habitat? *Bitácora, 10*(1), 124–135. http://www.revistas.unal.edu.co/index.php/bitacora/article/view/18717/19614 (accessed on 21 December 2021).

Cubillos-González, R. A., (2017). Design principles of social housing that is resilient to climate change | Principios para el diseño de vivienda social resiliente frente al cambio. In: Molar, O. M. E., (ed.), *The 20th Century Housing Challenges* (p. 222). Universidad autónoma de Cohauila. http://www.investigacionyposgrado.uadec.mx/libros/2017/2017LosRetosViviendaSigloXXI.pdf (accessed on 21 December 2021).

Cubillos-González, R. A., Novegil-gonzález-anleo, J. F., & Cortés-Cely, O. A., (2017). *Resilient and Efficient Territories in Bogotá.* https://www.dijuris.com/libro/territorios-resilientes-y-eficientes-en-bogota_36993 (accessed on 21 December 2021).

Cubillos-González, R. A., Trujillo, J., Cortés, C. O. A., Milena, R. C. M., & Villar, L. M. R., (2014). Habitability as a variable in the design of the building oriented towards sustainability. *Revista de Arquitectura, 16*, 114–125.

Cvetković, D., Nešović, A., & Terzić, I., (2021). Impact of people's behavior on the energy sustainability of the residential sector in emergency situations caused by COVID-19. *Energy and Buildings, 230*. https://doi.org/10.1016/j.enbuild.2020.110532.

D'alessandro, D., Gola, M., Appolloni, L., Dettori, M., Fara, G. M., Rebecchi, A., Settimo, G., & Capolongo, S., (2020). COVID-19 and living space challenge. Well-being and public health recommendations for a healthy, safe, and sustainable housing. *Acta Biomedica, 91*(1), 61–75. https://doi.org/10.23750/abm.v91i9-S.10115.

Del, G. V., De Paola, P., & Del, G. F. P., (2020). COVID-19 infects real estate markets: Short and mid-run effects on housing prices in Campania region (Italy). *Social Sciences, 9*(9). https://doi.org/10.3390/SOCSCI9070114.

DiMaggio, C., Klein, M., Berry, C., & Frangos, S., (2020). Black/African American Communities are at highest risk of COVID-19: spatial modeling of New York City ZIP Code-level testing results. *Annals of Epidemiology, 51*, 7–13. https://doi.org/10.1016/j.annepidem.2020.08.012.

Domínguez-amarillo, S., Fernández-agüera, J., Cesteros-garcía, S., & González-lezcano, R. A., (2020). Bad air can also kill: Residential indoor air quality and pollutant exposure risk during the COVID-19 crisis. *International Journal of Environmental Research and Public Health, 17*(19), 1–34. https://doi.org/10.3390/ijerph17197183.

Druta, O., Rogers, D., & Power, E., (2020). Holiday reading list for a post-COVID-19 housing system. *International Journal of Housing Policy, 20*(4), 467–473. https://doi.org/10.1080/19491247.2020.1838788.

Fazzini, M., Baresi, C., Bisci, C., Bna, C., Cecili, A., Giuliacci, A., Illuminati, S., Pregliasco, F., & Miccadei, E., (2020). Preliminary analysis of relationships between COVID-19 and climate, morphology, and urbanization in the Lombardy region (Northern Italy). *International Journal of Environmental Research and Public Health, 17*(19), 1–13. https://doi.org/10.3390/ijerph17196955.

Gilderbloom, J. H., & Meares, W. L., (2020). How inter-city rents are shaped by health considerations of pollution and walkability: A study of 146 mid-sized cities. *Journal of Urban Affairs*, 1–17. https://doi.org/10.1080/07352166.2020.1803751.

He, J., Dang, Y., Zhang, W., & Chen, L., (2020). Perception of urban public safety of floating population with higher education background: Evidence from Urban China. *International Journal of Environmental Research and Public Health, 17*, 8663.

Ikiz, E., Maclaren, V. W., Alfred, E., & Sivanesan, S., (2021). Impact of COVID-19 on household waste flows, diversion, and reuse: The case of multi-residential buildings in Toronto, Canada. *Resources, Conservation, and Recycling, 164*, 105111. https://doi.org/10.1016/j.resconrec.2020.105111.

Jankovic, L., (2020). Experiments with self-organized simulation of movement of infectious aerosols in buildings. *Sustainability (Switzerland), 12*, 5204.

Keenan, J. M., (2020). COVID, resilience, and the built environment. *Environment Systems and Decisions, 40*(2), 216–221. https://doi.org/10.1007/s10669-020-09773-0.

Lak, A., Asl, S. S., & Maher, A., (2020). Resilient urban form to pandemics: Lessons from COVID-19. *Medical Journal of the Islamic Republic of Iran, 34*(1), 1–9. https://doi.org/10.34171/mjiri.34.71.

López-Escamilla, Á., Herrera-limones, R., & León-Rodríguez, Á. L., (2020). Environmental comfort as a sustainable strategy for housing integration: The AURA 1.0 prototype for social housing. *Applied Sciences Article, 10*, 7734.

Maalsen, S., & Dowling, R., (2020). COVID-19 and the accelerating smart home. *Big Data and Society, 7*(2). https://doi.org/10.1177/2053951720938073.

Maalsen, S., Rogers, D., & Ross, L. P., (2020). Rent and crisis: Old housing problems require a new state of exception in Australia. *Dialogues in Human Geography, 10*(2), 225–229. https://doi.org/10.1177/2043820620933849.

MacKinnon, L., Socías, M. E., & Bardwell, G., (2020). COVID-19 and overdose prevention: Challenges and opportunities for clinical practice in housing settings. *Journal of Substance Abuse Treatment, 119*, 108153. https://doi.org/10.1016/j.jsat.2020.108153.

Mahendra, S., & Mittal, S., (2020). Improvised rental housing to make cities COVID safe in India. *Cities, 106*, 102922. https://doi.org/10.1016/j.cities.2020.102922.

Mericle, A. A., Sheridan, D., Howell, J., Braucht, G. S., Karriker-Jaffe, K., & Polcin, D. L., (2020). Sheltering in place and social distancing when the services provided are housing and social support: The COVID-19 health crisis and recovery housing. *Journal of Substance Abuse Treatment, 119*, 108094. https://doi.org/10.1016/j.jsat.2020.108094.

Mikolai, J., Keenan, K., & Kulu, H., (2020). Intersecting household-level health and socio-economic vulnerabilities and the COVID-19 crisis: An analysis from the UK. *SSM-Population Health, 12*, 100628. https://doi.org/10.1016/j.ssmph.2020.100628.

Moustafa, K., (2020). Make good use of big data: A home for everyone. *Cities, 107*. https://doi.org/10.1016/j.cities.2020.102903.

Nguyen, Q. C., Huang, Y., Kumar, A., Duan, H., Keralis, J. M., Dwivedi, P., Meng, H. W., et al., (2020). Using 164 million google street view images to derive built environment predictors of COVID-19 cases. *International Journal of Environmental Research and Public Health, 17*(17), 1–13. https://doi.org/10.3390/ijerph17176359.

Osborne, L. P., Cushing, D. F., & Washington, T. L., (2020). Changing greenspace in residential developments in an inner suburb of Brisbane, Australia. *Australian Planner*. https://doi.org/10.1080/07293682.2020.1824198.

Pombo, A., Luz, C., Rodrigues, L. P., Ferreira, C., & Cordovil, R., (2020). Correlates of children's physical activity during the COVID-19 confinement in Portugal. *Public Health, 189*, 14–19. https://doi.org/10.1016/j.puhe.2020.09.009.

Ralli, M., Cedola, C., Urbano, S., Morrone, A., & Ercoli, L., (2020). Homeless persons and migrants in precarious housing conditions and COVID-19 pandemic: peculiarities and prevention strategies. *European Review for Medical and Pharmacological Sciences, 24*(18), 9765–9767. https://doi.org/10.26355/eurrev_202009_23071.

Rogers, D., & Power, E., (2020). Housing policy and the COVID-19 pandemic: the importance of housing research during this health emergency. *International Journal of Housing Policy, 20*(2), 177–183. https://doi.org/10.1080/19491247.2020.1756599.

Rosenberg, A., Keene, D. E., Schlesinger, P., Groves, A. K., & Blankenship, K. M., (2020). COVID-19 and hidden housing vulnerabilities: Implications for health equity, New Haven, Connecticut. *AIDS and Behavior, 24*(7), 2007, 2008. https://doi.org/10.1007/s10461-020-02921-2.

Rubino, I., Coscia, C., & Curto, R., (2020). Identifying spatial relationships between built heritage resources and short-term rentals before the COVID-19 pandemic: Exploratory perspectives on sustainability issues. *Sustainability (Switzerland), 12*(11). https://doi.org/10.3390/su12114533.

Sacco, M. A., Caputo, F., Ricci, P., Sicilia, F., De Aloe, L., Bonetta, C. F., Cordasco, F., et al., (2020). The impact of the COVID-19 pandemic on domestic violence: The dark side of home isolation during quarantine. *The Medico-Legal Journal, 88*(2), 71–73. https://doi.org/10.1177/0025817220930553.

Saez, M., Tobias, A., & Barceló, M. A., (2020). Effects of long-term exposure to air pollutants on the spatial spread of COVID-19 in Catalonia, Spain. *Environmental Research, 191*. https://doi.org/10.1016/j.envres.2020.110177.

Sharifi, A., & Khavarian-Garmsir, A. R., (2020). The COVID-19 pandemic: Impacts on cities and major lessons for urban planning, design, and management. *Science of the Total Environment, 749*, 1–3. https://doi.org/10.1016/j.scitotenv.2020.142391

Sharma, S. V., Chuang, R. J., Rushing, M., Naylor, B., Ranjit, N., Pomeroy, M., & Markham, C., (2020). Social determinants of health-related needs during COVID-19 among low-income households with children. *Preventing Chronic Disease, 17*, E119. https://doi.org/10.5888/pcd17.200322.

Silva, D. S., (2020). COVID-19 in the public housing towers of Melbourne: upholding social justice when invoking precaution. *Australian and New Zealand Journal of Public Health, 44*(5), 430. https://doi.org/10.1111/1753-6405.13041.

Sliwa, K., & Yacoub, M., (2020). Catalyzing the response to NCDI Poverty at a time of COVID-19. *The Lancet, 396*(10256), 941–943. https://doi.org/10.1016/S0140-6736(20)31911-5.

Tokazhanov, G., Tleuken, A., Guney, M., Turkyilmaz, A., & Karaca, F., (2020). How is COVID-19 experience transforming sustainability requirements of residential buildings? A review. *Sustainability, 12*(20), 8732. https://www.mdpi.com/864674(accessed on 21 December 2021).

Van, E. N. J., & Waltman, L., (2019). VOSviewer. In: *Leiden University* (1.6.11). Centre for Science and Technology Studies, Leiden University, The Netherlands. http://www.vosviewer.com/(accessed on 21 December 2021).

Wu, Y., Yan, X., Zhao, S., Wang, J., Ran, J., Dong, D., Wang, M., Fung, H., Yeoh, E., & Chung, R. Y. N., (2020). Association of time to diagnosis with socioeconomic position and geographical accessibility to healthcare among symptomatic COVID-19 patients: A retrospective study in Hong Kong. *Health & Place, 66*, 102465. https://doi.org/10.1016/j.healthplace.2020.102465.

Xie, J., Luo, S., Furuya, K., & Sun, D., (2020). Urban parks as green buffers during the COVID-19 pandemic. *Sustainability (Switzerland)*, *12*(17), 1–17. https://doi.org/10.3390/SU12176751.

Zoğal, V., Domènech, A., & Emekli, G., (2020). Stay at (which) home: second homes during and after the COVID-19 pandemic. *Journal of Tourism Futures*, *2022*. https://doi.org/10.1108/JTF-06-2020-0090.

CHAPTER 5

Good Logistics Produces Healthy Spaces

JUAN FLAVIO MOLAR OROZCO

Industrial Engineering, National Technological Institute of Mexico/ Technological Campus of Ciudad Madero, Av. 1o. de Mayo s/n 89440, Ciudad Madero, Tamaulipas, México,
E-mail: juan.mo@cdmadero.tecnm.mx

5.1 INTRODUCTION TO LOGISTICS

Today, there are a lot of talks about logistics, but it is easy to know that this concept is not clear to many people. The concept of logistics and its characteristics are explained in a simple way down below.

One of the many definitions of logistics and the most complete one is the following:

> *Process of planning, implementing, and controlling the flow and storage of raw materials, semi-finished or finished products, and to handle related information within its process, from its origin to the place of consumption, with the purpose of satisfying the consumers requirements (Council of Logistics Management, 1985).*

Now, what does this mean?

It means that logistics is the gathering of all the activities that are carried out, within, and outside the company, which have to do with *moving and storing* its resources. The resources can be:

- **Tangibles and Intangibles:**
  - **Tangibles:** Materials, products, people, money.
  - **Intangibles:** Information, ideas.

Architecture for Health and Well-Being: A Sustainable Approach. Maria Eugenia Molar Orozco, PhD (Ed.)

- **Internals and Externals:**
    - **Internals:** Materials, products, or people of the company (workers).
    - **Eternals:** Materials, products, or people foreign to the company (clients, suppliers).

5.1.1 MOVE RESOURCES

The moving of the resources refers to the translation from its origin to an assigned destination, and all the activities involved and the necessary equipment to achieve that these resources arrive on time and in shape to the assigned destination; for example, moving raw materials inside of the productive process, move finished products all the way to its client, or move the clients inside of a system of service (Figure 5.1). It is important to be clear that we have to design the system more efficiently to move the resources searching to fulfill the commitment declared by the customer.

FIGURE 5.1 Public transport in England.
Source: Photograph of María Molar.

5.1.2 STORE RESOURCES

Storing resources is a complicated concept because the word store is related the majority of times with a warehouse, but from the logistics approach, it should be understood as the stage of the process where the resources must wait before they can be used or served.

In other words, storing does not refer to the action of putting the resource inside a warehouse, but to rather the process of conserving the resource in adequate conditions while waiting to be utilized or cared for.

For example:

- Under what conditions should food be preserved before it can be used to prepare a dish?
- What conditions are the ones suitable for the raw material of a product to be used?
- How should customers wait to be served? What type of queue should they do, or in what way are they served?
- Which is the electronic format or virtual that is more suitable to store the data and information?

It is important to undergo good planning of our storage system that allows resources to stay healthy during the wait time (Figure 5.2).

FIGURE 5.2 Lobby in a library.
Source: Photograph of María Molar.

In summary, Logistics is responsible for ensuring that nothing fails.

"When everything goes well, nobody notices. But if there is any complaint or discomfort on the part of the customer, is that logistics failed" —(Authors own phrase)

It is important to carry out good planning of our storage system that allows the resources to remain in good condition during the waiting time.

In summary, Logistics is itself responsible for ensuring that nothing fails. "When everything goes well, nobody notices. But if there is any complaint or discomfort on the part of the client, it is that the logistics failed." (Author's own phrase).

It is important to pay close attention to this phrase because we must understand that those activities related to the movement and storage of the company's resources, which are generally very common or simple activities, are sometimes not given the necessary importance to plan them well, and this leads to high-cost results and customer complaints. Today the competitive environment that companies live in has brought us to a point where no company can afford to lose customers. "When everything goes more than well, and the client recognizes it, then we generate a good experience for the client, and this helps us achieve customer loyalty."

5.1.3 LOYALTY

And what is loyalty?

Loyalty: It is a marketing term that seeks for the customer to be faithful to our product, service, or brand, where we can ensure that they do not change us for another provider. Today there are so many customer options that we must do everything possible to retain our customers.

Currently, customers have become more subjective, and the difference between products or services is not the product itself, but rather the experience of the service process or the use of the product that generates the customer's decision to buy it again. Use it or recommend it. The key to being competitive is not in making a good product and designing a good service, but in generating a good experience in the customer, based on their emotions and feelings when using our product or service (Figure 5.3).

FIGURE 5.3 Happy faces.
Source: Author's photograph.

5.2 CUSTOMER-FOCUSED LOGISTICS

It has been said that logistics is in charge of moving and storing resources within production systems, and when this is applied in material things, it is with the aim of meeting customer requirements and seeking to do so at the lowest possible cost. But when this is done in people, we must do it with the aim of generating a good experience for the customer and also at the lowest possible cost.

The last part is really difficult, and we will see an example:

Let us suppose that they must move 50 boxes of a product (bottles of wine, canned food, cereal boxes, etc.), from an origin and to a programmed destination. Only the required vehicle must be selected according to the characteristics demanded by the product and that it has the adequate capacity, then draw the route that is shortest, and with that, it could be said that good logistics would be achieved to guarantee meeting the customer's requirements at a lower cost (Figure 5.4).

Suppose that 50 people must be transferred from an origin to a programmed destination; in this case, there are more variables to take into account:

- Vehicle type;
- Type of seats;
- Vehicle temperature;
- Kind of music;
- Way of driving;
- Vehicle speed;
- Shortest route;
- Homogeneity of passengers;
- Process of attention to the arrival of passengers;
- Process of attention to the departure of passengers.

FIGURE 5.4 Satisfaction of public transport in Spain.
Source: Photograph of María Molar.

In this case, the best effort could be made to comply with all these variables as best as possible, but if at least one of the 50 passengers did not have a good experience in any of them, the objective was no longer met.

It could be that he got cold, or he did not like the music, that the seat was uncomfortable for him, the sunlight was giving him all the way, perhaps he did not like the driver's way of driving because he was very slow or fast depending on his perception, or because he was touched by a person who bothered him all the way.

It could be just one thing which can cause a bad experience for that customer. Those details to which we do not give importance in our contact with the client may be the ones that cause null customer loyalty (Figure 5.5).

FIGURE 5.5 Bad experience.
Source: Author's photograph.

5.3 GOALS OF LOGISTICS

The two main goals of logistics are:

- Provide an acceptable level of service to customers; and
- Operate a logistics system that allows to satisfy in general, the requirements of the clients.

It can be seen that the first goal indicates that the level of service can vary depending on each client. In other words, each client can have an acceptable level of the service they want. And this can only be achieved if we have a good understanding of our customers.

It is very important to be clear that in order to achieve effective logistics management, we must first know very well what the requirements of our customers are. And these requirements can vary from client to client. In a few words, the first goal refers to the fact that it is necessary to know what are those requirements that the customer is looking for from our product, be it a good or service. To the extent that we have information about our clients in relation to their needs, tastes, and preferences, it would be necessary to adjust our processes to maintain that level of acceptance that we need to achieve. In this sense, it is important to develop mechanisms to collect information directly from the client. Some companies use questionnaires or quick interviews; other companies use mechanisms such as opinion or complaint boxes; in some products, a telephone number or web address is indicated so that the customer can offer their opinions or suggestions. However, all these actions depend on the client being willing to collaborate, and the percentage of clients is very low.

Today it is necessary to start capitalizing on the experiences or mistakes of the company, that is, be very attentive and observe what happens every day within our company. We must learn to perceive the reactions of our customers during the contact processes with our product or service. In addition, we must be very attentive about the information that is handled in social networks in relation to our product or service, because the subjective comments of a customer can influence the buying process of others.

For example, on one occasion, a film was being shown in a chain of cinemas nationwide, but before the film, a short film was shown. The fact is that the short film lasted about 30 minutes and this began to cause that through the social networks people complained about the length of the short film, after 1 week the chain of cinemas eliminated the screening of the short film.

From this event, we can see the following:

- For this company, customer feedback is important;
- The company uses all possible means to be in contact with its customers (Figure 5.6);
- The company adapts its processes to customer satisfaction;
- The client is the one who defines the processes;
- For the company, the client is the main actor.

We can learn to know our clients to the extent that we can be sensitive to their reactions and needs.

FIGURE 5.6 Digital medium.
Source: Author's photograph.

On the other hand, Goal 2 tells us that it is necessary to develop processes that help us meet Goal 1, that is, once we know the acceptable level of service that the client wants, we must design and effectively implement the administrative processes, productive, and service necessary to meet the customer.

These processes must include the procedures, personnel, equipment, and infrastructure necessary to achieve customer satisfaction and at the same time translate into a good experience for the customer (Figure 5.7).

FIGURE 5.7 Infrastructure and space.
Source: Photograph of María Molar.

5.4 LOGISTICS IN SERVICE

Whether the customer buys a tangible good or receives a service, there is always a moment of contact with the customer in distribution logistics.

5.4.1 DISTRIBUTION LOGISTICS

It is the process through which the product is delivered to the customer.

This process is becoming more important today because it is where the customer's shopping experience is achieved. Originally the aspects that the client required of this process were the following:

- Delivery on time (Figure 5.8);
- Correct product;
- Product in good condition;
- Product without defects.

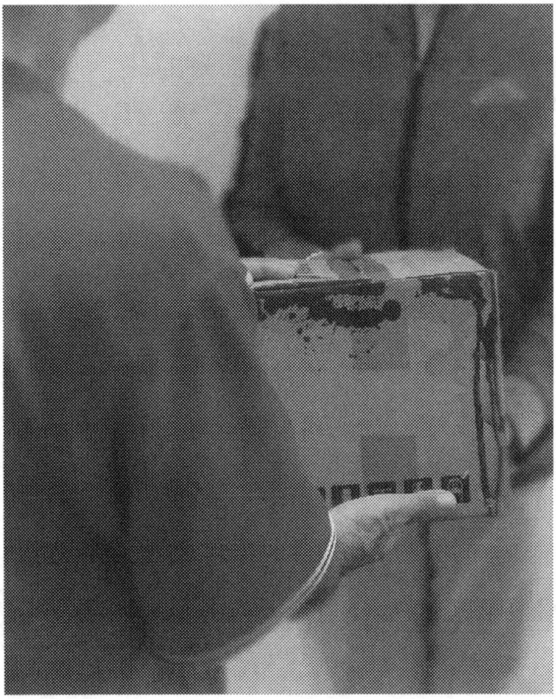

FIGURE 5.8 Delivery on time.
Source: Render by Hartz (2020); adapted from: RoseBox on Unsplash.

By complying with these aspects, an acceptable level of the client was achieved. But today, this level of the client has changed, and now other aspects are required, such as:

- Kindness and good treatment (Figure 5.9);
- Ease of access to the business (parking, escalators, elevator, ramps for the disabled);
- Good atmosphere (air-conditioned premises, lighting, background music);
- Adequate furniture (comfortable furniture) (Figure 5.10);
- Adequate infrastructure (comfortable waiting areas);
- Technological infrastructure (communication and computing systems, GPS, RFID);
- Security systems;
- Hygiene and cleaning systems.

FIGURE 5.9 Rules of kindness.
Source: Author's photograph.

FIGURE 5.10 Space and furniture.
Source: Author's photograph.

In this sense, companies must make a bigger effort every day to improve their processes and facilities in order to maintain an acceptable level of customer service.

To the extent that they are clear that their service processes, as well as their facilities are important to ensure that the customer has a good shopping experience, then they can develop strategic plans that help them take care of these details.

5.5 LOGISTICS AND INFRASTRUCTURE

As you have seen, today, consumers are evaluating their shopping experiences more to determine the product or service they will use. Increasingly, we see that the customer or consumer returns to that supplier who managed to produce a good experience in him.

There are many cases where the experience has more value for the customer than the product itself, we return for the place, the environment, the treatment or simply for the moment we will spend there. Companies that are understanding this begin to work by modifying their processes and facilities to achieve customer loyalty.

We can see how some advertisements from leading companies speak more about the service processes and facilities than about the product itself, because they know that the client is looking for options that help them obtain good life experiences.

Customers value an experience more than product quality, they know that product quality, today, can be offered by anyone, but good service is not easy to achieve. In this sense, the infrastructure plays an important role because the customer on many occasions prefers to make their purchases in places that have easy and convenient access.

For example, businesses that are easy to move, those that have parking or those that are protected from the elements of the weather (sun, rain, cold).

They also look for places with pleasant environments, with pleasant temperatures (neither cold nor hot), cleaning, and hygiene services, and a good security system. They also look for places with new technologies that offer them fast and comfortable services, those places where they have to wait, at least, for it to be in a pleasant and comfortable way.

For example, there are customer service offices where the service system is done by means of a number that is given to the customer as he arrives, and the customer waits his turn in a seating area instead of standing (Figure 5.11).

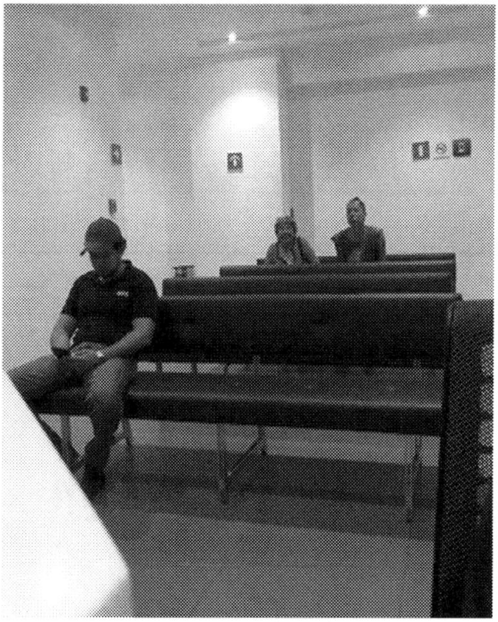

FIGURE 5.11 Waiting space for clients.
Source: Author's photograph.

Logistics and Architecture must work hand in hand to design functional spaces that help achieve good experiences for clients and consumers (Figure 5.12).

It is important that the design of our logistics processes and the design of our stores are integrated to achieve the objective of producing good experiences for our customers and consumers (Figure 5.13).

FIGURE 5.12 Waiting space in a movie theater.
Source: Author's photograph.

FIGURE 5.13 Customer support.
Source: Author's photograph.

Therefore, it is important that the professionals in charge of the design of the facilities related to the client, know the characteristics of the service and also the level of acceptance, so that they design the necessary spaces to generate a good experience for the client (Figure 5.14).

FIGURE 5.14 Public space of a shopping plaza.
Source: Author's photograph.

5.6 RELATIONSHIP BETWEEN HEALTH AND LOGISTICS

5.6.1 HEALTH

How does WHO define health?

> *Health is a state of complete physical, mental, and social well-being, and not just the absence of diseases or illnesses.*

The quote comes from the Preamble to the Constitution of the World Health Organization (WHO), which was adopted by the International Sanitary Conference, held in New York from June 19 to July 22, 1946, signed on July 22, 1946, by the representatives of 61 states (Official Records of the World Health Organization, No. 2, p. 100), and entered into force on April 7, 1948. The definition has not been modified since 1948 (Figure 5.15).

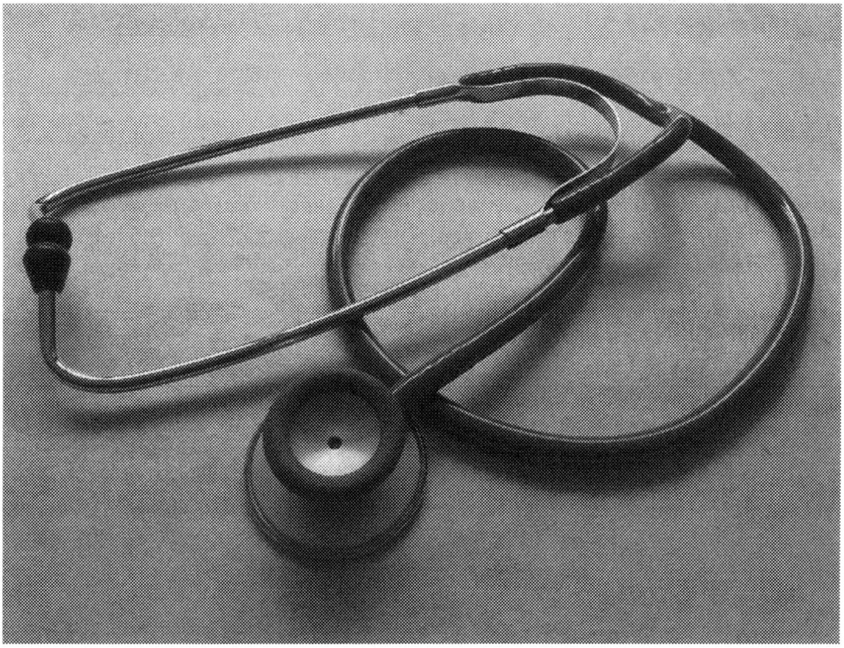

FIGURE 5.15 Device of a doctor.
Source: Author's photograph.

According to the previous definition, a person's health has a direct relationship with their environment, which includes physical and environmental conditions of the place where they are, in addition to their form of physical and mental interaction with other people.

From the above we can see that in the aspect of physical and environmental facilities is where we can find the relationship that it has with logistics.

On the other hand, there is also the fact that diseases most of the time are the result of infections caused by transmission between people due to the proximity between them.

It is important to see that the distance between people is the result of the spaces that are used during daily processes within the facilities in which we carry out work, recreation, rest, food, social activities, celebrations, etc. (Figure 5.16).

FIGURE 5.16 Entertainment space at La Pedrera, Barcelona.
Source: Photograph of María Molar.

At this time where we are, a density very high of the people of the world, the spaces we have, to carry out our daily activities, are increasingly reduced. We must also take into account that means of public transport are another transmission factor between people, and that thousands and even millions of people move daily in large cities, causing in most cases physical contacts that cause health problems through external agents.

If we analyze the above a little, we can realize that the main logistics activities, transport, and storage, can become important factors to prevent the spread of diseases to some extent and have, as a result, the power to maintain good health.

Today we must begin to realize the importance of carrying out a planning process for the design and implementation of transportation systems for people (Figure 5.17), looking for spaces that favor the prevention of contagion among passengers, as well as in the waiting systems in services (Figure 5.18), where clients must remain before being served.

FIGURE 5.17 Adequate spaces in transport, Spain.
Source: Photograph of María Molar.

FIGURE 5.18 Ample waiting area at Barcelona airport.
Source: Photograph of María Molar.

The foregoing with the aim of achieving the protection of people's health. We can identify some of the variables that are related to these logistics systems and that to some extent can affect people's health.

For example, in a transportation system there are the following variables:

- Temperature;
- Noise;
- The smell;
- The ventilation system;
- The movement;
- Safety;
- The lighting;
- The comfort;
- The physical space;
- Stress.

All these variables can affect, to some extent, the physical, mental, or even social health of people, and for this reason the logistical design of transport systems is necessary.

In the case of Waiting systems for clients to be served, we can also see that these same variables can affect the health of clients.

5.6.1.1 TEMPERATURE

One of the indicators of good health in people is their body temperature, which can be affected by external or internal factors.

In a transport system and also in a waiting system, there must be a mechanism that controls the environmental temperature that favors a good body temperature of people, with the aim of not affecting their health (Figure 5.19).

It is also necessary for the temperature to satisfy to some extent an environment of comfort and convenience for the user.

When people are in some type of transport or in places that have low temperatures and do not have some type of clothing and special accessories to protect themselves from the cold, this can cause health problems related to their respiratory system. For this reason, the planning process is very important within the logistics processes, as well as in the design of physical spaces.

FIGURE 5.19 Space in a cableway.
Source: Photograph of María Molar.

5.6.1.2 THE NOISE

When people are in places, either within a transport or in a physical space, where there are unpleasant or annoying sounds, their health can be affected. Noise can damage people's hearing and, in some cases, cause physical and psychological disorders in the human system.

For these reasons, it is important to design transport systems that have mechanisms to reduce the noise caused during their use, and also that they also have, where possible, environmental sound systems that help reduce the stress and anxiety of people during the transfer time.

On the other hand, also in spaces where people have to wait to be served, they must have ambient sound systems designed to help reduce noise and provide music that reduces the feeling of time that passes while people are being served (Figure 5.20).

5.6.1.3 SMELL

Smell is one of the senses that through the olfactory system generates a pleasant or unpleasant sensation in people, due to the breathing of a mixture of gases, vapors, and dust, which are found in the environment.

FIGURE 5.20 Space away from noise at La Pedrera, Barcelona.
Source: Photograph of María Molar.

Within a means of transport, such as in a waiting place, there is always the possibility of having pleasant or unpleasant smells, which can also cause discomfort in the environment, resulting in some cases, in effects on the quality of life or the people's health.

So, in certain cases they can be considered as environmental pollution. The process by means of which a pleasant aroma, of a bad smell, could be identified, is very difficult because being one of the senses of man, it carries a large percentage of subjectivity. In other words, what for someone may be a bad smell, for another person is not perceived that way.

The relationship between smells and health is that there are some smells that can affect people's health and can cause dizziness and headaches, as well as respiratory disorders and even psychological disorders. Bad odors can have really harmful effects on health.

In this sense, it is important that transport systems and waiting places have appropriate mechanisms to reduce or eliminate bad odors (Figure 5.21).

FIGURE 5.21 Inside the subway of Barcelona.
Source: Photograph of María Molar.

5.6.1.4 THE VENTILATION SYSTEM

A Ventilation System is a mechanism that serves to renew air within a closed or open area. There are several ways to do it, from natural ways or with manual or automated mechanical technologies.

Regardless of how it is done, the purpose is that living beings can have fresh and clean air during their stay in the place (Figure 5.22).

These systems have a direct relationship with the odor variable, discussed above, because they must be used during the design process of types of transport or waiting places for people. Seeking the protection of the health of users and clients (Figure 5.23).

FIGURE 5.22 Space of a bar in Barcelona.
Source: Photograph of María Molar.

FIGURE 5.23 Bathroom space.
Source: Photograph of María Molar.

5.6.1.5 MOVEMENT

When a body changes position, in a space, for a certain time it can be affected due to the magnitude of force that has been exerted on it. This applies to both things and people, for this reason, it is necessary for transport systems to have safety mechanisms that protect products and people from sudden movements that cause serious physical injury.

The physical and mental health of people can be affected due to movements during their transfer from the point of origin to the point of destination.

It is important to carry out a process of planning, implementation, and control of the transport of people, which guarantees the protection of their physical and mental health during the journey, through the use of mechanisms and procedures designed as a result of an analysis of data on the variables involved in a certain task (Figure 5.24).

If we say that the transport activity is a logistical task, then logistics will be responsible for taking care of people's health.

FIGURE 5.24 Inside a train Bagon in Spain.
Source: Photograph of María Molar.

5.6.1.6 SAFETY

It is the situation where conditions and dangers can be controlled so that physical, mental, or psychological damage does not occur with the aim of preserving the well-being and health of a community or an individual.

It is very common to see insecurity everywhere; we can see how people move without taking into account their safety and much less that of others. There are also many places that do not have preventive security systems, where at all times the People can see their physical, psychological, or mental health damaged.

A large percentage of health damages caused by accidents are the result of not having prevention measures that help to minimize the injuries caused by the event, in this sense we can see that to comply with these prevention measures there are responsibilities for both people who design the equipment or spaces, but also the people who use them. Talking about safety is talking about the subject of culture, safety is a cultural process where the individual must first know how to maintain their safety and then must carry out the actions required to preserve their health.

When we see someone performing unsafe actions that may cause damage to their health and we make them see the situation, it is very likely that they have a pretext to justify their actions, even knowing that they put their health at risk, in these cases it is very common that think that accidents are for others and not for him.

Within logistics, it is very important that in the planning stage the use of mechanisms and procedures focused on accident prevention is always taken into account in order to safeguard people's health in general (Figure 5.25).

5.6.1.7 LIGHTING

It is the effect of illuminating or putting light on physical things or in a certain space. For people to be able to see something clearly, they first require that there is a lighting mechanism, and to the extent that this mechanism provides an adequate amount of light to objects, it will be the result of the clarity of the things to be observed.

On the other hand, having bad lighting can cause health problems in people; from fatigue, headaches, vision problems, stress, among others. Also, bad lighting can cause accidents and affect both health or quality of life in people. Within logistics activities such as transport and waiting processes in physical spaces, there is a need for lighting mechanisms that help prevent health problems in people who operate or use these activities (Figure 5.26).

FIGURE 5.25 View of a train platform.
Source: Photograph of María Molar.

Currently, there are techniques, methods, and technological developments that help the professionals in charge of the design of transport equipment and physical spaces to carry out their work, seeking that their users can perform their tasks in a faster, more comfortable way and safe.

Today there are many ways to achieve enlightenment, but it depends on each situation and purpose to determine which is the most appropriate, but in the end, it is very important not to lose sight of the fact that people's health should be a priority.

5.6.1.8 COMFORT

This variable is subjective and depends on the characteristics of the people: Age, Culture, Sex, Weight, etc. The designers of the types of transport or waiting places try to create conditions that are acceptable to users and must also take their health into account.

FIGURE 5.26 Lighting in the central of the Barcelona subway.
Source: Photograph of María Molar.

When a person is in a state of comfort, we must assume that her basic human needs for relief, calm, and well-being are satisfied.

Comfort is a result of people's environment and that includes several factors, which together can achieve well-being, for this reason, it is very important to analyze the factors that make up the environment, so that comfort is can achieved.

In the following graphs, we can see the comfort factors and parameters (Figures 5.27–5.31).

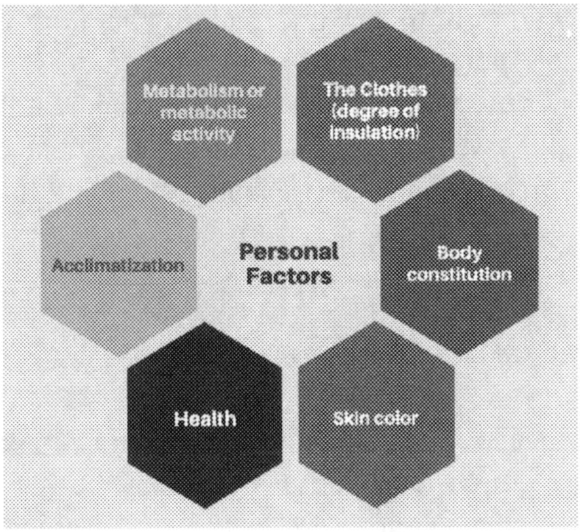

FIGURE 5.27 Personal factors.
Source: Author's graphic.

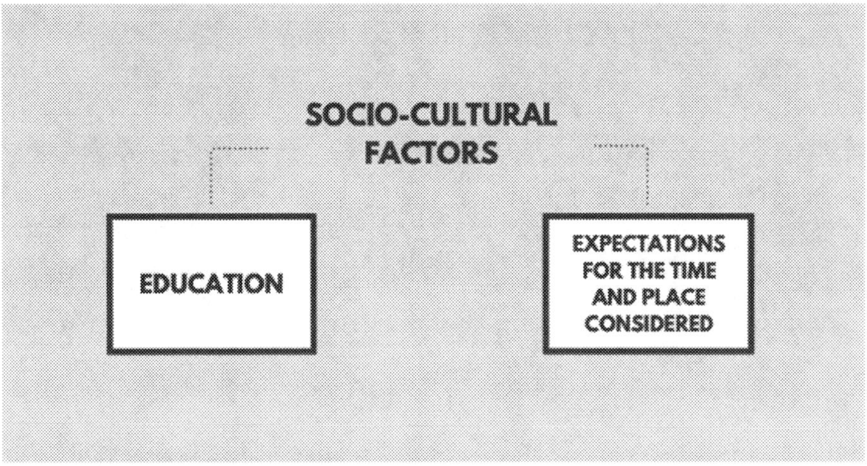

FIGURE 5.28 Socio-cultural factors.
Source: Author's graphic.

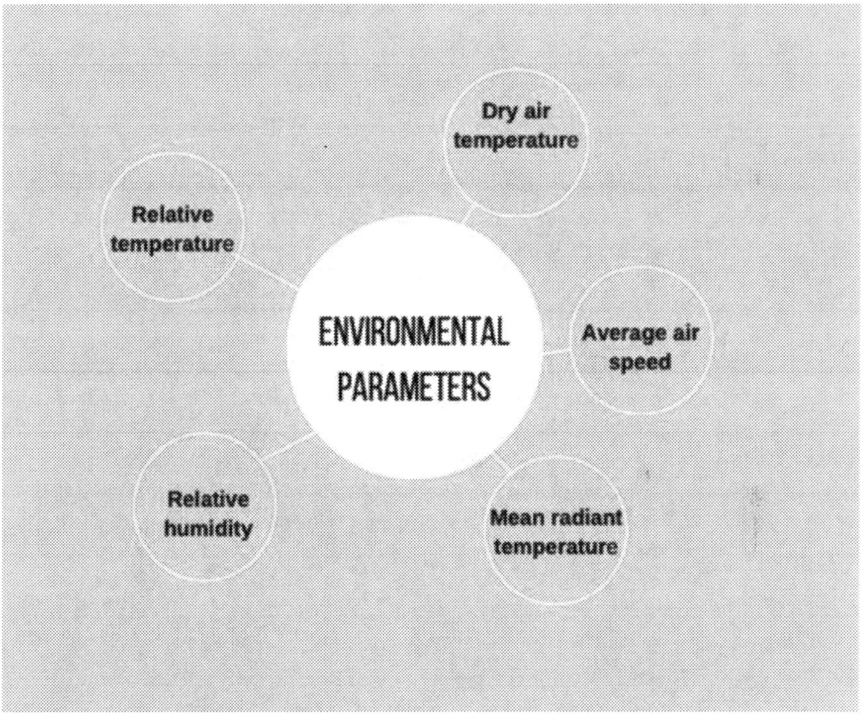

FIGURE 5.29 Environmental parameters.
Source: Author's graphic.

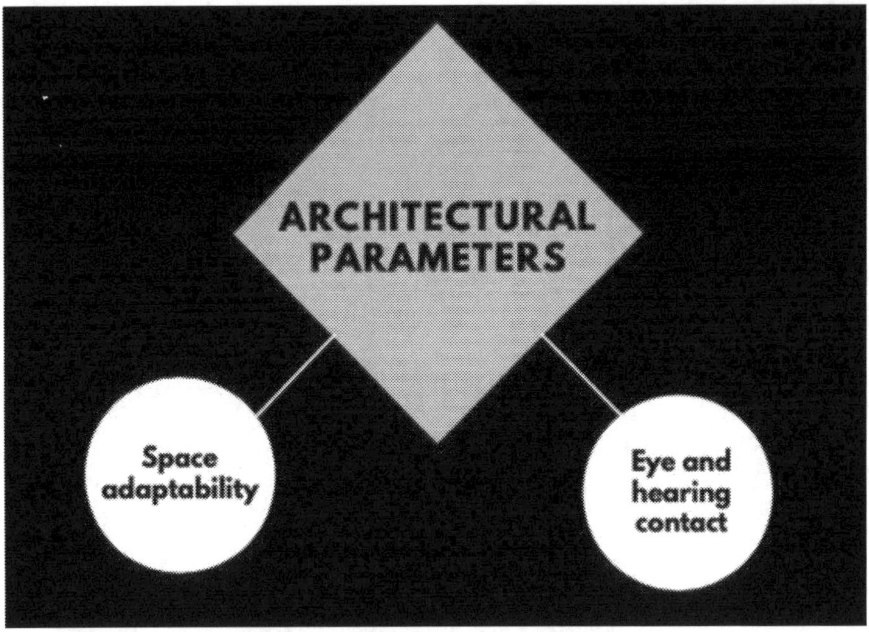

FIGURE 5.30 Architectural parameters.
Source: Author's graphic.

FIGURE 5.31 Space with natural lighting.
Source: Photograph of María Molar.

5.6.1.9 *PHYSICAL SPACE*

In order to understand the concept of this variable, we are going to say that it refers to what is called the built environment, which is the set of physical elements that make up a specific place and that was designed and developed by the man.

We can say that living beings are always in some physical space and that we travel through them at all times, this means that we live in 4-dimensional spaces where time is a main factor for this variable.

When we talk about physical spaces, most of the time we focus on the physical and environmental characteristics of the place; dimensions, objects, lighting, ventilation mechanisms, comfort, etc. But it is also necessary to take into account the time factor, on the understanding that we are going to spend some time in that place or physical space (Figure 5.32).

When we can see this, then there is the possibility of building transport equipment and waiting places where people can maintain good health, as a result of the time spent within these physical places.

It is important to bear in mind that the sensation of time in people is relative, it often depends on the characteristics of the built environment, and that that sensation of time can be a cause that affects people's health and well-being. In addition to the fact that this physical space requires to have been built with the necessary mechanisms to guarantee and safeguard the physical, mental, and psychological health of people (Figure 5.33).

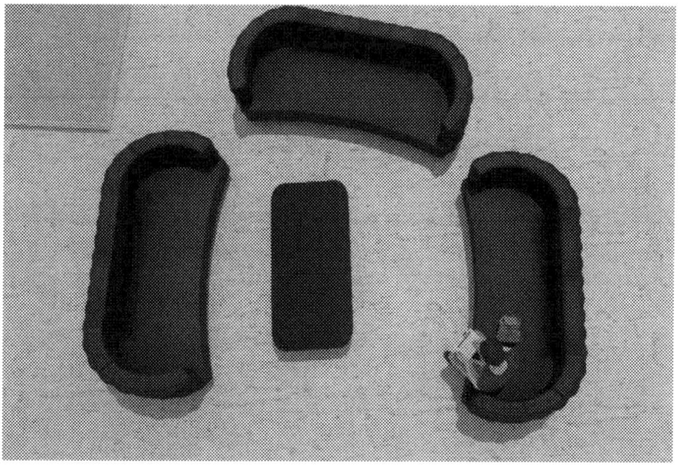

FIGURE 5.32 Space to rest o relax.
Source: Render by Hartz (2020); adapted from: John Salvino on Unsplash.

FIGURE 5.33 Stay in a transport.
Source: Render by Hartz (2020); adapted from: Vincent Guth on Unsplash.

5.6.1.10 STRESS

Environmental stress. Lack of adjustments between the properties of the physical environment and the needs of the individual (Karminoff and Proshansky, 1982).

Psychological stress It is the process of responding to demanding, overstimulating, or threatening situations for the individual's well-being (Evans and Cohen, 1987).

People who carry out logistics activities (transport and storage) can be affected by moments of stress of any of these types, caused by the situations of their work, as well as those people who are users or clients of these processes (Figure 5.34).

This leads us to take into account that the design of transport equipment, workplaces, and waiting rooms for people, require analysis processes with respect to these factors that can cause damage to health (Figure 5.35).

5.6.2 CASE STUDY; COVID-19 PANDEMIC

In December 2019, in the City of Wuhan, capital of the Hubei province, in the People's Republic of China, the coronavirus disease 2019 (COVID-19)

(Figure 5.36) caused by the SARS-CoV2 virus was identified for the first time (severe acute respiratory syndrome coronavirus type 2), when reporting cases of a group of sick people with an unknown type of pneumonia. It was determined that most of those infected had some kind of relationship with workers at the wholesale seafood market in Wuhan city.

FIGURE 5.34 Person with stress at work.
Source: Render by Hartz (2020); adapted from: Christian Erfurt on Unsplash.

FIGURE 5.35 Person with stress in transportation.
Source: Render by Hartz (2020); adapted from: Lily Banse on Unsplash.

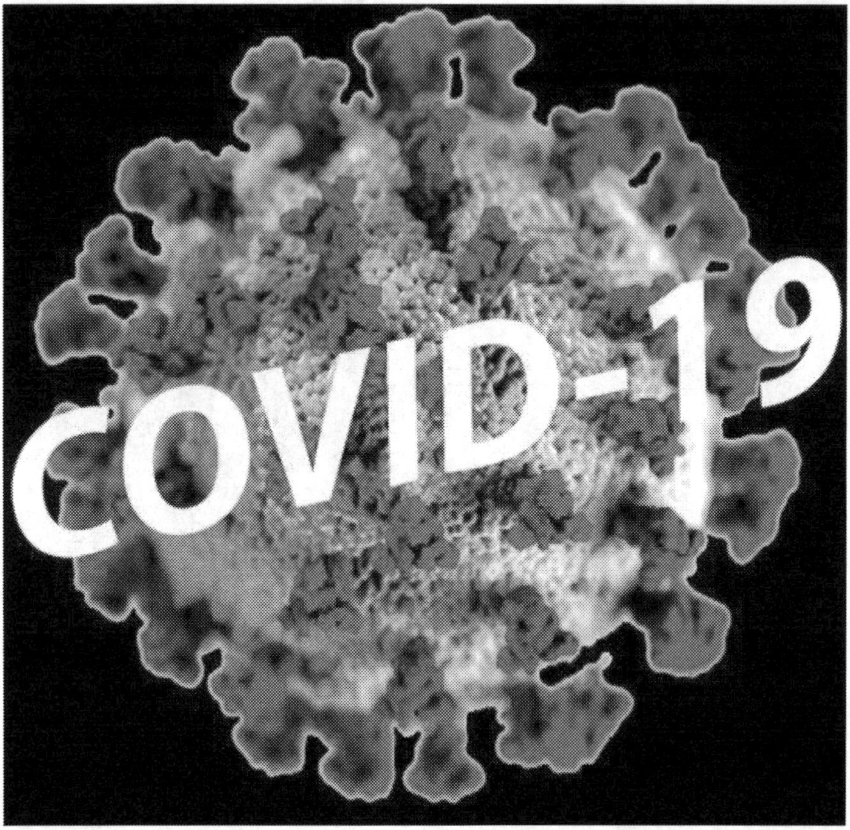

FIGURE 5.36 COVID-19.
Source: Render by Hartz (2020); adapted from: Martín Sánchez on Unsplash.

A Pandemic is when there is the contagion of an infectious disease throughout a very large geographic area, in this case when the infection of people reaches several countries. This COVID-19 pandemic managed to reach most cities around the world in a few months, and for the month of October 2020, there was information that there were more than 40.6 million people affected by the virus in 220 countries and territories in the world (Figure 5.37). (It is worth mentioning that as of the date this chapter was written, there was still no vaccine and the number of infected still did not stop).

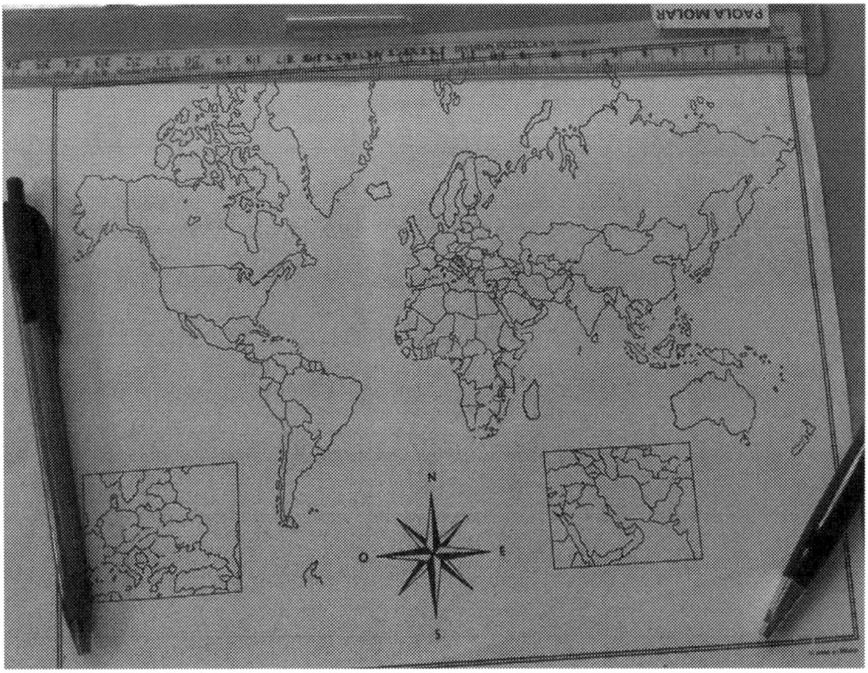

FIGURE 5.37 World map.
Source: Author's photograph.

One of the main factors of the great speed and spread of the virus is the current level of globalization in the world, as well as the development in the logistics processes of transporting people and products worldwide.

The cause and the solution for this type of pandemic depend on good logistical management, in addition to the fact that this pandemic required a change in the daily lives of all people. Once the first official contagion occurred somewhere, the activities of all inhabitants were affected by the necessary prevention processes to contain and reduce the number of cases of infected people.

5.6.2.1 STAYING IN HOME

The main recommendation to avoid contagion was "Stay Home," but what did Staying in Home mean? (Figure 5.38).

FIGURE 5.38 Stay home message in Spanish.
Source: Author's photograph.

The action is very simple, do not leave the house at all, but most people require activities that are done outside the home, for example (Figure 5.39):

- Job;
- Study;
- Fun;

- Exercise;
- Food;
- Provision of supplies.

FIGURE 5.39 Work activities, study, and fun.
Source: Author's photograph.

At first it was had the idea that the necessary time of permanence in the house would be short, and that in a couple of months, it would be possible to return to normality, but as soon as it was seen that the number of infections did not stop to increase, then measures were taken so that activities could be carried out from home (Figure 40).

TOTAL CASES OVER TIME AROUND THE WORLD	
JANUARY 2020	9,826
FEBRUARY 2020	85,403
MARCH 2020	750,867
APRIL 2020	3'090,445
MAY 2020	5'934,936
JUNE 2020	10'185,374
JULY 2020	17'106,007
AUGUST 2020	25'118,689
SEPTEMBER 2020	33'561,077
OCTOBER 2020	40'696,053

FIGURE 5.40 Infected people.

Source: Own elaboration based on Google Noticias (2020).

In this sense, a radical change began in the way of working and studying for each family, now activities had to be carried out from home. Fortunately, advances in information and communication technologies helped us to adapt more easily to the new situation.

The spaces of the house were now the new classrooms and the new work offices, the individuals had to adapt and adapt to the new study and work conditions (Figures 5.41–5.43). What used to be only an area for food or rest, now it would also become an area for remote communication, study, and work. Work meetings and social gatherings had to be done remotely through video conferencing applications, which began to have unexpected demands and had to expand their capabilities and rapidly develop improvements in their handling (Figures 5.44 and 5.45).

FIGURE 5.41 Change of use of spaces.
Source: Author's photograph.

FIGURE 5.42 Change of use of spaces.
Source: Author's photograph.

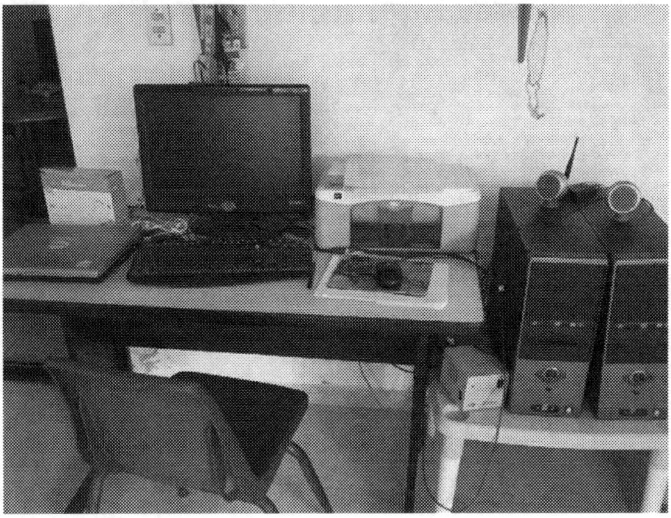

FIGURE 5.43 Change of use of spaces.
Source: Author's photograph.

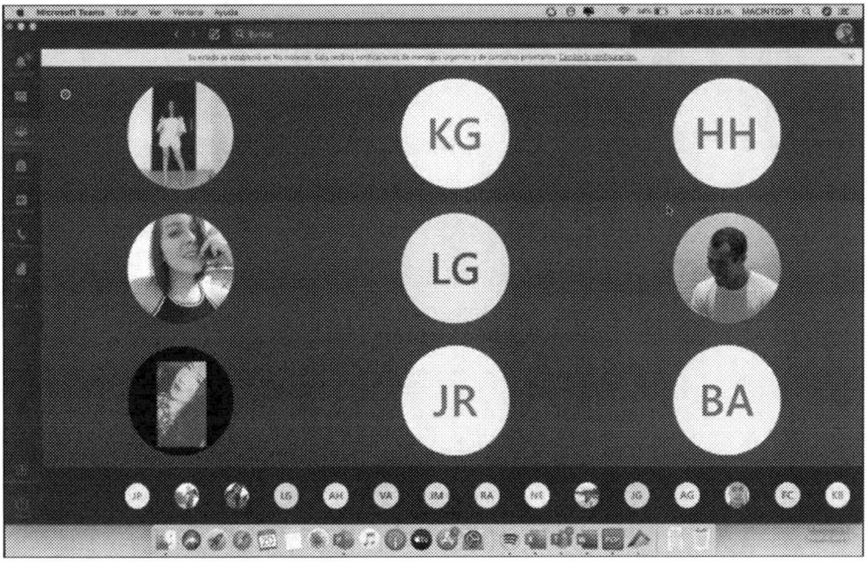

FIGURE 5.44 Virtual classes.
Source: Author's photograph.

FIGURE 5.45 Virtual meetings.
Source: Author's photograph.

The physical distancing between people and the use of masks was also another prevention measure for the spread of the disease; for this reason, special protocols were developed to be able to carry out activities in public and private places (Figures 5.46–5.49).

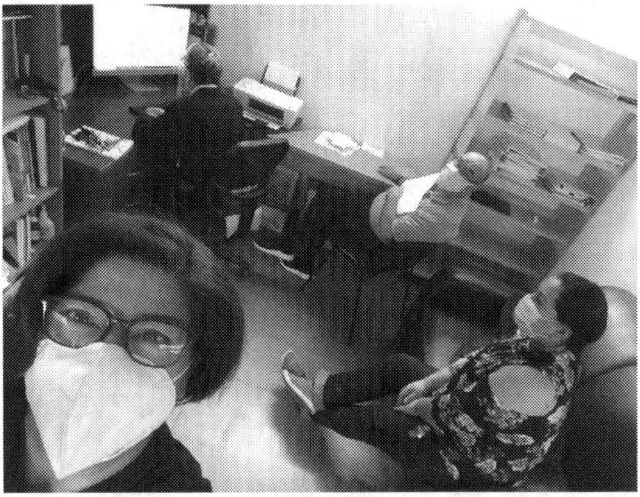

FIGURE 5.46 Use of protocols in workspaces.
Source: Photograph of María Molar.

FIGURES 5.47 Protocols in public spaces with signage.
Source: Photograph of María Molar.

FIGURE 5.48 Protocols in public spaces, protection barriers.
Source: Photograph of María Molar.

FIGURE 5.49 Protocols in banks.
Source: Author's photograph.

At the beginning, the governments ordered the temporary closure of places of public coexistence:

- Schools;
- Government offices;
- Parks and recreation centers;
- Restaurants;
- Malls;
- Non-essential activity companies.

This was where logistics activities took on an important role for the prevention of contagion from people. In other words, the logistics activities of Transport and Store, had to be carried out effectively and efficiently in order to maintain good health in people.

Some companies that did not handle the home service had to start using it in order to continue carrying out their product sales, more products began to be demanded through electronic mechanisms and the development of new applications for this purpose emerged. But the logistics distribution activity had to continue, now with the protocols of healthy distance between suppliers and customers. Logistics activities became important in their execution for the prevention of contagion (Figures 5.50 and 5.51).

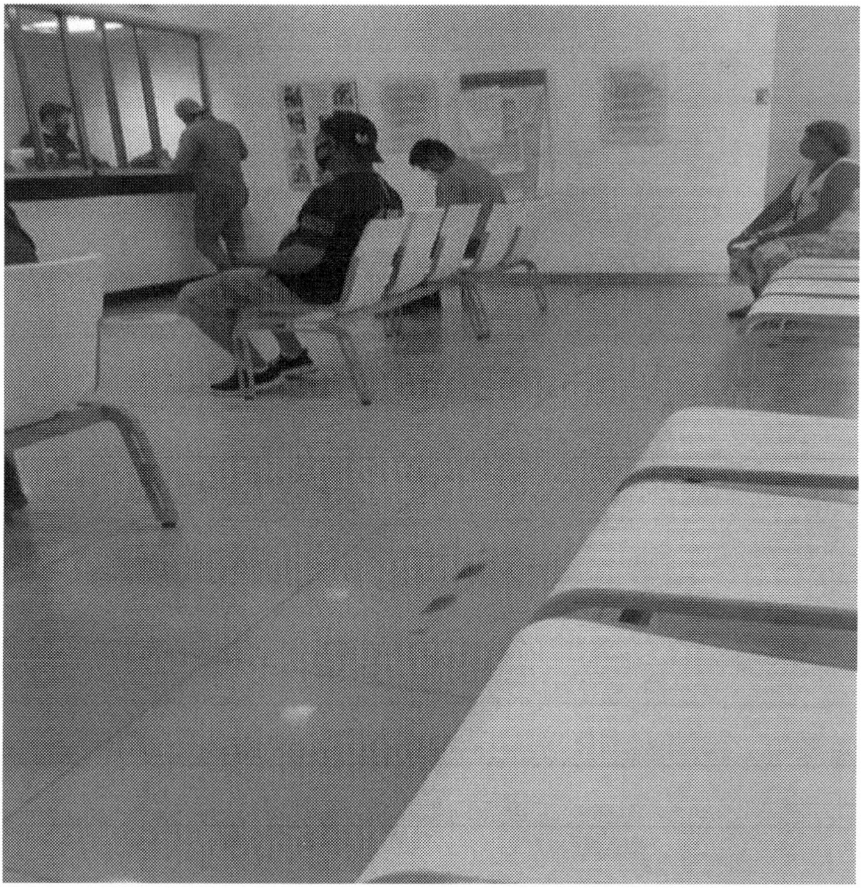

FIGURE 5.50 Signs to maintain social distance.
Source: Author's photograph.

In this case, if we talk about the storage activity seen as the space where it is expected to be attended, the protocols indicated the minimum necessary distances between people, as well as the mandatory use of a mask and alcohol in hand gel, also the use of sanitizing mats at the entrance of businesses or public offices (Figures 5.52–5.54).

FIGURE 5.51 Signs to maintain social distance.
Source: Author's photograph.

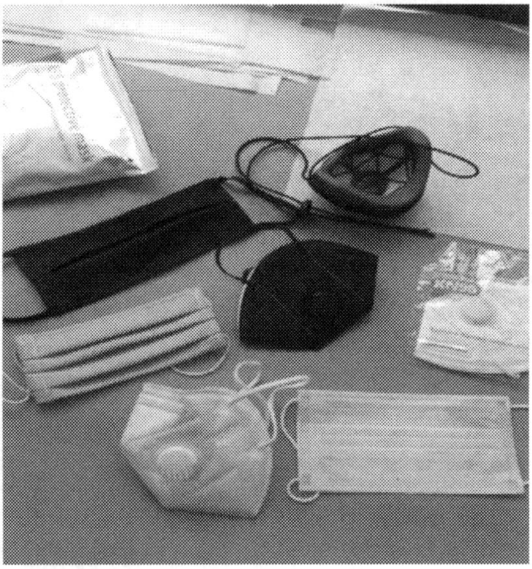

FIGURE 5.52 Personal protection devices.
Source: Author's photograph.

FIGURE 5.53 Personal protection devices.
Source: Photograph of María Molar.

FIGURE 5.54 Sanitizing mat, personal protection devices.
Source: Photograph of María Molar.

All this situation of the COVID-19 pandemic has drastically changed the way of life of people around the world, and companies have had to redesign themselves very quickly in order to survive, because in addition to the health problem, there is also the economic problem in companies and families.

Many companies had to stop providing service and close their businesses to the public or decrease their customer service capabilities, others have had to redefine their distribution processes seeking to comply with healthy distance protocols and reinforcing their transportation and delivery processes.

The preservation of health and the prevention of contagions became the main factors for the planning, implementation, and control of the flow and storage of resources in companies and in people's homes.

The traditional ways for people to move from one place to another, and also the traditional ways for people to wait to be served, in any process of marketing and service, have changed to what is called the new normal of lifetime.

5.6.2.2 CONCLUSION OF THE CASE

Logistics activities are always necessary in the daily life of companies, families, and people and we must carry them out always seeking to be effective and efficient for the production of healthy spaces. If we can do this well in the face of the COVID-19 pandemic, we can have good results.

KEYWORDS

- **atmosphere**
- **logistics**
- **loyalty**
- **security systems**
- **technological infrastructure**

REFERENCES

Ballou, R. H., (2004). *Logistic: Supply Chain Management*. Pearson, Prentice-Hall.

Evans, G. W., & Cohen, S., (1987). Environmental stress. In: Stokols, D., & Altman, I., (eds.), *Handbook of Environmental Psychology* (Vol. 1, pp. 571–610). New York: John Wiley & Sons.

Frohlich, & Katherine, L., (2013). *Neighborhood Structure and Health Promotion* Springer, Boston, MA.

Google News, (2020). *Coronavirus COVID-19*. Recovered from: https://news.google.com/covid19/map?hl=es419&mid=%2Fm%2F02j71&gl=MX&ceid=MX%3Aes-419 (accessed on 21 December 2021).

http://eadic.com/wp-content/uploads/2013/09/Tema-3-Confort-Ambiental.pdf (accessed on 21 December 2021).

https://es.wikipedia.org/wiki/COVID-19 (accessed on 21 December 2021).

Kaminoff, R., & Proshansky, H., (1982). Stress as a consequence of the urban physical environment. In: Goldberger, L., & Breznitz, S., (eds.), *Handbook of Stress: Theoretical and Clinical Aspects* (pp. 380–409). New York: MacMillan Publishing.

Restrepo, H. E., & Málaga, H., (2001). *Health Promotion: How to Build Healthy Life*. Pan American Health Org.

Zikmund, W. G., McLeod, Jr. R., & Gilbert, F. W., (2004). *CRM: Customer Relationship Management.* CECSA.

Index

D

E

G

H

I

Q

R

T

Printed in the United States
by Baker & Taylor Publisher Services